CAMBRIDGE STUDIES
IN MATHEMATICAL BIOLOGY: 1

Editors
C. CANNINGS
Department of Probability and Statistics, University of Sheffield
F. HOPPENSTEADT
Department of Mathematics, University of Utah

EVOLUTION IN AGE-STRUCTURED POPULATIONS

BRIAN CHARLESWORTH
School of Biological Sciences, University of Sussex

Evolution in age-structured populations

CAMBRIDGE UNIVERSITY PRESS

Cambridge

London   New York   New Rochelle

Melbourne   Sydney

Published by the Press Syndicate of the University of Cambridge
The Pitt Building, Trumpington Street, Cambridge CB2 1RP
32 East 57th Street, New York, NY 10022, USA
296 Beaconsfield Parade, Middle Park, Melbourne 3206, Australia

First published 1980

Printed in Great Britain by
J. W. Arrowsmith Ltd., Bristol BS3 2NT

British Library Cataloguing in Publication Data

Charlesworth, Brian
Evolution in age-structured populations.—
(Cambridge studies in mathematical biology).

1. Population genetics
2. Age
3. Evolution
I. Title   II. Series
575   QH455   79-41452

ISBN 0 521 23045 4 hard covers
ISBN 0 521 29786 9 paperback

TO MY PARENTS

CONTENTS

PREFACE

The aim of this book is to summarise our current knowledge of evolutionary theory as applied to age-structured populations, i.e. populations whose members are not born into distinct generations, and where fertility and survival are functions of age. Although theoretical treatments of this subject date back to the beginnings of population genetics, its growth has been relatively slow until the last decade or so, when important contributions have been made by workers in several different areas. Population geneticists have subjected many species with age-structured populations to experiments and observations, *Drosophila* and man being the two best-known examples. A clear understanding of the implications of age-structure for such processes as selection and genetic drift is therefore necessary for a proper interpretation of experimental findings. Similarly, animal and plant breeders often wish to make predictions about the effects of artificial selection regimes imposed on age-structured populations. An understanding of the effects of natural selection on life-history phenomena, such as ageing and the timing of reproductive effort, also requires a quantitative theory of the effects of selection in relation to age. For these reasons, a survey of the present sort seems timely.

My policy has been to develop the mathematical treatment in some detail, so that the reader can see for himself the logic behind the results presented. I have deliberately kept the mathematics to an elementary level, so that anybody with a knowledge of basic calculus and matrix algebra should have no difficulty in following the derivations. A knowledge of elementary genetics is assumed. More advanced mathematical techniques are essential to a full treatment of certain topics, such as the application of bifurcation theory to the study of density-dependent populations, the treatment of temporally varying survival rates and fecundities, and the use of optimal control theory in life-history evolution. I have contented myself with citations of what I hope are the major papers on these subjects.

I have made some effort to discuss the implications of the theoretical results for empirical biologists, and many examples of the application of the theory to real data are given or quoted. I hope that the material is arranged in such a way that readers who are not interested in the details of the mathematical derivations will be able to identify, and make sense of, the major conclusions. I have tried to give a broad coverage of what I consider to be the most important results in the field. Inevitably, much work is only referred to or treated briefly, regardless of its intrinsic importance, and I trust my colleagues will not feel I have unduly emphasised my own work at their expense. By no means all the problems in this field have been solved, and I hope I have indicated clearly where work remains to be done.

Chapter 1 is a general survey of the basic mathematical theory of the demography and ecology of age-structured populations, and introduces many of the concepts and notation used in later chapters. Chapter 2 deals with such questions as the approach to Hardy–Weinberg equilibrium and the effects of genetic drift. It also contains a brief account of the problem of calculating the expected frequencies of consanguineous matings in a finite population. Chapters 3 and 4 cover the theory of gene frequency change and equilibrium under selection. Among other things, they contain discussions of the effects of demographic changes on gene frequencies at polymorphic loci, the appropriateness of various fitness measures, and the effects of selection on a quantititiave character. Chapter 5 is concerned with life-history evolution in general, with particular emphasis on the evolution of senescence and on the reproductive effort model of optimal life-histories.

Acknowledgements

This book was largely written between October 1977 and September 1978, while I was a visiting scientist in the Laboratory of Environmental Mutagenesis, National Institute of Environmental Health Sciences, North Carolina, U.S.A. I am greatly indebted to the Institute for its generosity in offering me the opportunity of spending a year there, and to its staff for their many kindnesses throughout my stay. I am particularly grateful to my host, Charles Langley, who provided a highly stimulating intellectual environment, and to Denise D. Crawford for her efficient typing of the bulk of the manuscript.

I wish to thank the following people for their patience in reading and commenting on the manuscript: C. Cannings, D. Charlesworth, J. F. Crow, F. Hoppensteadt, R. Lande, C. H. Langley, J. Maynard Smith, T. Nagylaki, M. R. Rose, and W. M. Schaffer. They (particularly Thomas Nagylaki, who criticised Chapters 1, 2, 3 and part of 4 with great care) have saved me from many errors, ambiguities and lapses from English. I have not always taken their advice, and the remaining defects are, of course, entirely my responsibility.

Finally, I must express my gratitude to Chris Cannings for suggesting that I write this book and for his encouragement during its preparation.

1

Models of age-structured populations

1.1 Introduction

The populations of many species of plants and animals are *age structured*; if such a population is investigated at a given point in time, it is found to consist of a set of individuals who were born over a range of past times, and whose fecundities and probabilities of future survival depend on their age. The human population is an obvious example. A full ecological description of the state of such a population should therefore take into account the relative numbers of individuals of different ages (the age-structure of the population), as well as its overall size. Similarly, a description of the genetic composition of the population should contain a classification of individuals by age as well as genotype. The traditional theories of ecology and population genetics have, however, usually ignored the problem of age-structure, and have assumed that populations can be treated as homogeneous with respect to age. This is clearly legitimate in dealing with organisms with discrete generations, such as annual plants, where only individuals born in the same season are capable of interbreeding. Many species of biological importance have overlapping generations, however, and their fecundities and survival rates are rarely completely independent of age, so that their age-structure should be taken into account in any realistic model of their behaviour.

There has recently been considerable interest in extending ecological and population genetics models to include age-structure as a variable. Evolutionary theorists have also become interested in the role of natural selection in moulding the way in which fecundity and survival depend on age. The purpose of this book is to give an account of the major results which have been obtained by applying ecological and demographic models of age-structured populations to population genetics and evolutionary theory. Before going on to

1

this topic, it is desirable to develop the basic models and concepts which are needed, and this is the subject of the present chapter.

We are primarily concerned in this chapter with the construction and analysis of mathematical models which enable us to handle changes in the size and age-structure of biological populations. This is the subject-matter of mathematical demography, a science which dates from the work of Euler (1760), although its main development was initiated by Lotka and his associates in this century (Sharpe and Lotka, 1911; Lotka, 1925). More comprehensive treatments than the one given here may be found in the books by Keyfitz (1968), Pollard (1973) or Henry (1976). It is assumed throughout this chapter that populations are genetically homogeneous, or at least constant in their genetic make-up, so that individuals can legitimately be classified solely by sex and age. It is also assumed that the populations with which we are concerned are so large that stochastic effects can be ignored, and that numbers of individuals can be adequately represented by continuous variables. The concepts and notation introduced in this chapter are essential for an understanding of later chapters, where we discuss the problems associated with describing and predicting evolutionary changes in age-structured populations.

We begin with a discussion of methods for describing the state of an age-structured population at a given time, and then show how the state at one time can be related to that at previous times. This is followed by a detailed analysis of the process of change in state of a population which lives in an unchanging environment, and whose numbers are not restricted by density-dependent factors resulting from competition for limited resources, such as food. Although most species of animals or plants only rarely, and for short periods, find themselves in such an environment, the concepts which emerge from the study of this case are fundamental to an understanding of the dynamics of age-structured populations in general. We then discuss briefly the effects of temporal fluctuations in the survival and fecundity parameters of a density-independent population, and go on to consider the problem of density-dependent factors. The chapter concludes with a brief consideration of the dynamics of systems of several, non-interbreeding populations.

1.2   The description of age-structured populations

1.2.1   *Populations with discrete age-classes*

Probably the simplest type of biological population which can usefully be described as age-structured is one which is characteristic of many temperate-zone species of vertebrates and perennial plants. Individuals may survive over many years, but reproduction is limited to one season of the year. In a given breeding season, individuals who were born in several different, earlier, breeding seasons may be reproductively active, so that it is impossible to assign individuals to separate generations. The composition of the population when censused at the beginning of the breeding season in a given year t can be described in terms of the numbers of males and females falling into *age-classes* 1, 2, 3, etc. These correspond to individuals who were born 1, 2, 3, ... years previously. The number of females aged x at the beginning of the breeding season in year t can be written as $n(x, t)$, and the corresponding number of males as $n^*(x, t)$. Our problem is to relate these numbers to the state of the population in the preceding year, $t-1$. There are basically two methods of doing this; one is to formulate the problem in terms of matrix algebra, and the other is to use difference equations of high order. These two methods yield equivalent results, and it is solely a matter of technical convenience which is to be used in approaching a particular problem. We shall therefore consider both of them in this chapter. We start by developing the matrix formulation for a number of cases, and then go on to discuss the difference equation approach.

The Leslie matrix.   The population at the start of the breeding season in year t is composed partly of individuals who were present at the corresponding period of year $t-1$, and partly of individuals who were born during the breeding season of that year. The latter individuals constitute age-class 1 for year t. We can write $P(x, t)$ and $P^*(x, t)$ for the probabilities of survival over one year for females and males, respectively, who are present in age-class x at the start of the breeding season in year t. These probabilities are usually functions of age, and will be referred to in what follows as the *age-specific survival probabilities*. They are also written as functions of time, to indicate that in general they are expected to change with time in response to changes in factors of the physical environment, or to changes in the numbers of individuals in the population. The probability that a female aged x in year t dies during the year can be written as $Q(x, t) = 1 - P(x, t)$; a corresponding quantity $Q^*(x, t) = 1 - P^*(x, t)$ can obviously be defined for males. These are the *age-specific probabilities of death*.

Using this notation, we have, for age-classes other than the first, the relations

$$n(x, t) = n(x - 1, t - 1)P(x - 1, t - 1) \qquad (x > 1) \qquad (1.1a)$$

$$n^*(x, t) = n^*(x - 1, t - 1)P^*(x - 1, t - 1) \qquad (x > 1) \qquad (1.1b)$$

It is more difficult to construct a realistic model of the way in which individuals in the first age-class are generated; in general this would require a detailed specification of the probabilities of mating events between pairs of individuals of all possible ages, and the fecundities of these matings. Fortunately, it is usually reasonable to assume that the fecundity of females of a given age is independent of the age of their mates, and that there are sufficient males available to fertilise the females. *Age-specific fecundities* can therefore be assigned to females; these are independent of the age-composition of the population. (This assumption is standard in demographic theory and is usually referred to as the *demographic dominance* of females.) Let $M(x, t)$ be the expected number of fertile eggs produced in year t by a female aged x at the start of the breeding season. Let the proportion of females among the offspring resulting from those eggs be $a(x, t)$; this is the *primary sex-ratio* for offspring of mothers aged x at time t. The expected number of female zygotes produced by a female aged x in year t is thus

$$m(x, t) = M(x, t)a(x, t) \tag{1.2}$$

In order to compute the number of individuals in age-class 1 at the start of the breeding season in a given year, we need to specify the probability of survival of a zygote over one year. For simplicity, we assume here that this probability, which is denoted by $P(0, t)$ for a female zygote produced in year t, is independent of the age of the parents. The net expected contribution from a female aged x in year t to the population of females aged 1 in year $t+1$ is thus

$$f(x, t) = m(x, t)P(0, t) \tag{1.3}$$

This may be called the *net fecundity* for the female concerned.

The assumption that the survival of an individual over the first year of life is independent of parental age is probably quite often violated in practice, especially for species with parental care, where senescence may affect the ability of parents to look after their young. For example, Perrins and Moss (1974) have shown that the rate of hatching of eggs of the great tit (*Parus major*), in English woodland, declines with maternal age. This problem may be avoided by censusing the individuals born in year t at a time when they are old enough for their future survival to be unaffected by parental age. $m(x, t)$ is then the number of female zygotes weighted by the probability of survival to this stage, and $P(0, t)$ is the probability of survival from this stage to the beginning of the breeding season in year $t+1$.

We are now in a position to relate the state of the population at two successive times. There is generally a upper limit to the age at which reproduction is possible or, if there is no such limit, so few individuals survive beyond a certain age that they can safely be disregarded. Let the last age-class which is taken into account have the index d. Similarly, there is a lower limit to the age of reproduction, $x = b$ say ($b \geqq 1$). We obtain from the above definitions the relation

$$n(1, t) = \sum_{x=b}^{d} n(x, t-1)f(x, t-1) \tag{1.4}$$

A full description of the population of immature and reproductively active females at time t is provided by the d-dimensional row vector $\boldsymbol{n}(t)$, whose components are $n(1, t)$, $n(2, t)$, ..., $n(d, t)$. (Throughout this book a row vector consisting of n elements $(x_1, x_2, \ldots x_n)$ is written as $\boldsymbol{x}$, and the corresponding column vector as the transposed vector $\boldsymbol{x}^{\mathrm{T}}$.) Equations $(1.1a)$ and (1.4) can be combined compactly in the matrix equation

$$\boldsymbol{n}(t) = \boldsymbol{n}(t-1)\boldsymbol{L}(t-1) \tag{1.5}$$

where $L(t-1)$ is the $d \times d$ matrix defined by the expression

$$
L(t-1) = \begin{bmatrix}
f(1, t-1) & P(1, t-1) & 0 & \ldots & 0 \\
f(2, t-1) & 0 & P(2, t-1) & \ldots & 0 \\
\vdots & \vdots & \vdots & & \vdots \\
f(d-1, t-1) & 0 & 0 & \ldots P(d-1, t-1) \\
f(d, t-1) & 0 & 0 & \ldots & 0
\end{bmatrix}
$$

(1.6)

All the elements of L are zero except for the off-diagonal elements $P(1, t-1), P(2, t-1), \ldots P(d-1, t-1)$, and some of the elements of the first column, $f(x, t-1)$. A matrix such as L is referred to as a *population projection matrix* or *Leslie matrix*, after P. H. Leslie, whose papers were largely influential in promoting its widespread use in ecology and demography (Leslie, 1945, 1948). This method was also suggested independently by Bernadelli (1941) and Lewis (1942). A matrix which, like the Leslie matrix, has elements which are positive or zero is called a *non-negative* matrix. Such matrices have special properties, which facilitate the analysis of the population process described by equation (1.5) (Seneta, 1973). These will be discussed later, in sections 1.3.1 and 1.4.1.

In most biological applications we are interested mainly in the population of pre-reproductive and reproductive individuals, so that the vector $n(t)$ provides an adequate description of the female population. If for some special purpose it is necessary to take into account individuals who are past reproductive age, it is easy to calculate their numbers using equation (1.1a). We have

$$
n(x, t) = n(d, t+d-x) \prod_{y=1}^{x-d} P(x-y, t-y) \quad (x > d)
$$

The male population. There are several ways to proceed if one wishes to include the male population in the analysis. One method is to define an age-specific fecundity function $M^*(x, t)$ for males, in terms of the number of young which a male aged x is expected to father in year t. If the frequency of males among these offspring is $1 - a^*(x, t)$, then the expected number of male progeny contributed by a male aged x in year t is

$$m^*(x, t) = M^*(x, t)[1 - a^*(x, t)] \qquad (1.7)$$

The net contribution of this male to the population of males in age-class 1 in year $t + 1$ can be obtained by weighting this fecundity by the probability of survival of a male zygote over the year, $P^*(0, t)$. By analogy with equation (1.3), this is

$$f^*(x, t) = m^*(x, t)P^*(0, t) \quad (x \geqq b^*) \qquad (1.8)$$

where b^* is the age at first reproduction for males. If the effective upper age-limit to male fertility is d^* (which is not necessarily equal to d), the male population at time t can be described by the d^*-dimensional row vector $n^*(t)$, with components $n^*(1, t)$, $n^*(2, t), \ldots n^*(d^*, t)$. A $d^* \times d^*$ Leslie matrix, L^*, can be constructed for the male population in the same way as for the female population in equation (1.6), giving the matrix equation

$$n^*(t) = n^*(t - 1)L^*(t - 1) \qquad (1.9)$$

This approach is not very useful from the point of view of studying the population process, since, with the assumption of female demographic dominance, the basic male fecundities $M^*(x, t)$ must depend in general on the age-composition of the female population. This can change in time, even if environmental conditions are constant, resulting in changes in the male fecundities (which are constrained by the condition that every individual has both a male and a female parent). A more useful method is to attribute all offspring to their female parents. The net contribution of male offspring in year $t + 1$ from a female aged x in year t is thus

$$f'(x, t) = M(x, t)P^*(0, t)[1 - a(x, t)] \qquad (1.10)$$

and so we obtain

$$n^*(1, t) = \sum_{x=b}^{d} n(x, t-1)f'(x, t-1) \qquad (1.11)$$

$n(t)$ and $n^*(t)$ can be combined into a single vector of dimension $d + d^*$, $[n(t), n^*(t)]$, which gives a complete description of the population at time t. From equations (1.1), (1.5) and (1.11), we have

$$[\boldsymbol{n}(t), \boldsymbol{n}^*(t)] = [\boldsymbol{n}(t-1), \boldsymbol{n}^*(t-1)]\left[\begin{array}{c|c} \boldsymbol{L}(t-1) & \boldsymbol{A}(t-1) \\ \hline \boldsymbol{0} & \boldsymbol{B}(t-1) \end{array}\right] \tag{1.12}$$

where $\boldsymbol{L}$ is the Leslie matrix defined by equation (1.6), $\boldsymbol{0}$ is a $d^* \times d$ matrix of zero elements; $\boldsymbol{A}$ is a $d \times d^*$ matrix all of whose elements are zero, apart from the first column, which consists of $f'(1, t-1)$, $f'(2, t-1), \ldots f'(d, t-1)$; $\boldsymbol{B}$ is a $d^* \times d^*$ matrix whose only non-zero elements are the off-diagonal elements $b_{x,x+1} = P^*(x, t-1)$, $x = 1, 2, \ldots d^* - 1$.

The approach embodied in equation (1.12) was introduced by Goodman (1969) and provides the simplest reasonably realistic method of handling an age-structured population with two sexes. For accounts of more complex approaches, the reader should consult Pollard (1973, Chapter 7) and Parlett (1972).

In species where reproduction is purely asexual, or where individuals are monoecious or hermaphroditic, as in many plants, the problems of describing a two-sex population clearly do not arise. The population process is completely characterised by a single equation similar to (1.5), except that we no longer have to discount male offspring in calculating net fecundities. The $f(x, t-1)$ are therefore equal to $M(x, t-1)P(0, t-1)$ in this case, and the vector $\boldsymbol{n}(t)$ refers to the whole population.

Plants with seed dormancy. In plant demography, however, the description of the population process is considerably complicated by the fact that seeds may persist for several years in the soil before germinating. In treating hermaphrodite plants that have seed dormancy, it is thus necessary to take into account the existence of a population of seeds of various ages, as well as the population of plants themselves. This topic is discussed in some detail by Harper and White (1974), Sarukhan and Gadgil (1974) and Harper (1977). Here we merely sketch the way in which one can describe the essentials of the population process, by extending the matrix formulation developed above.

It is convenient to treat both the seed and plant populations as being censused at the beginning of the flowering season in a given year, following the time of germination. We write $n'(1, t)$ for the number of seeds produced in the flowering season of year $t-1$ and which survived to year t without germinating, $n'(2, t)$ for the number produced in year $t-2$ which survived to t, etc. If $d'+1$ is the maximum number of years over which a seed can survive without germinating, the seed population is represented by the d'-dimensional row vector $n'(t) = [n'(1, t), n'(2, t), \ldots n'(d', t)]$. Seeds aged more than d' years can be ignored, since they contribute nothing to the adult population, and hence nothing to the future reproduction of the population. It is assumed that $d' \geqq 1$, so that seeds can survive more than one year; otherwise, it is unnecessary to treat the seed population as a distinct entity. Let $P'(x, t)$, for $x = 1, 2, \ldots d'-1$, be the probability that a seed aged x at the beginning of the season in year t does *not* germinate *and* survives for a year. We also define $P'(0, t)$ as the probability that a seed produced in year t survives to year $t+1$ without germinating. Similarly, $P''(x, t)$, for $x = 1, 2, \ldots d'$, is the probability that a seed aged x in year t survives to year $t+1$ *and then germinates*. $P''(0, t)$ is the probability that a seed produced in year t survives to year $t+1$ and germinates. The population of plants in year t is represented by the d-dimensional vector $n(t) = [n(1, t), n(2, t), \ldots n(d, t)]$; $n(1, t)$ is the number of plants present at the beginning of year t which result from newly germinated seeds, $n(2, t)$ is the number of plants which result from seeds which germinated in year $t-1$, etc. We write $P(x, t)$, with $x = 1, 2, \ldots d-1$, for the probability that a plant aged x in year t survives to year $t+1$ (where x is one plus the number of years since it germinated). Similarly, $M(x, t)$ is the expected number of seeds that a plant aged x in year t produces in that year. It is convenient to define two net seed fecundities, for non-germinating and germinating seeds, respectively, by the relations

$$f'(x, t) = M(x, t)P'(0, t) \tag{1.13a}$$

and

$$f''(x, t) = M(x, t)P''(0, t) \tag{1.13b}$$

Using these definitions, we obtain the following matrix equation for the population process

$$[n'(t), n(t)] = [n'(t-1), n(t-1)] \left[\begin{array}{c|c} A(t-1) & B(t-1) \\ \hline C(t-1) & D(t-1) \end{array} \right] \tag{1.14}$$

where A is a $d' \times d'$ matrix whose elements are all zero, except for the off-diagonals $a_{x,x+1} = P'(x, t-1)$; B is a $d' \times d$ matrix whose only non-zero elements are those of the first column, $P''(1, t-1)$, $P''(2, t-1)$, etc.; C is a $d \times d'$ matrix whose only non-zero elements are those of the first column, $f'(1, t-1)$, $f'(2, t-1)$, etc.; D is a $d \times d$ matrix whose only non-zero elements are the off-diagonals $d_{x,x+1} = P(x, t-1)$, and the elements of the first column, $f''(1, t-1)$, $f''(2, t-1)$, etc.

The difference equation approach. An alternative way of representing the population processes described above is by using high-order difference equations rather than matrix algebra. As will be seen in later chapters, this method often has certain advantages for population genetics problems. In this section, the procedure for the female population, corresponding to pp. 3–6 above, will be presented. The most convenient variable, with this method, for describing the state of the population in a given year t is the number of female zygotes produced in the breeding season of that year, $B_f(t)$. If $t \geqq d$, none of these zygotes will have been produced by individuals who were present at the start of the breeding season in some initial year, when t is taken as 0. We write $l(x, t)$ for the probability that a female survives from conception in year $t - x$ to age x at the beginning of the breeding season in year t. From the definitions presented above, on pp. 3–4, we have

$$l(x, t) = \prod_{y=1}^{x} P(x - y, t - y) \quad (x \geqq 1) \tag{1.15}$$

It is useful for some purposes to define $l(0, t)$ as equal to unity. These definitions imply that $l(x, t)$ is a decreasing function of x; if the survival probabilities are all less than unity, as is generally the case in practice, $l(x, t)$ is a strictly decreasing function of x. It will be referred to here as the *survival function.*

It is also convenient to have an expression for the net expectation of female offspring at age x, attributable to a female born at time $t - x$; this is given by the relation

$$k(x, t) = l(x, t)m(x, t) \tag{1.16}$$

This quantity is often referred to as the *reproductive function.*

Using these definitions, it is easily seen that, for $t \geq d$, we have

$$B_f(t) = \sum_{x=b}^{d} B_f(t-x)k(x, t) \tag{1.17}$$

For times such that $t < d$, some of the females who were present in the initial population are still reproductively active, and their contributions to $B_f(t)$ must be taken into account. An individual aged $x - t$ initially $(x \geq t+1)$ has a probability of survival to time t (when it will be aged x) of

$$\tilde{l}(x, t) = \prod_{y=1}^{t} P(x-y, t-y)$$

The total contribution of such individuals to $B_f(t)$ is thus

$$g(t) = \sum_{x=t+1}^{d} n(x-t, 0)\tilde{l}(x, t)m(x, t) \qquad (t < d) \tag{1.18}$$

For $t \geq d$, $g(t)$ can be defined as zero, so that equations (1.17) and (1.18) can be condensed into a single formula

$$B_f(t) = g(t) + \sum_{x=1}^{t} B_f(t-x)k(x, t) \tag{1.19}$$

Note that $k(x, t)$ is zero for $x > d$ and $x < b$.

The number of females in age-class x at time $t(t \geq x)$ is easily computed from $B_f(t-x)$ by the formula

$$n(x, t) = B_f(t-x)l(x, t) \tag{1.20a}$$

The definition of $B_f(t)$ as the number of female *zygotes* is purely a matter of convenience; any other stage in the first year of life may be chosen for this purpose, provided that $m(x, t)$ is re-defined so as to discount the fraction of zygotes which die before reaching the chosen stage. There must obviously also be a corresponding modification in $l(x, t)$. Finally, we may note that, if the primary sex-ratio is constant and independent of parental age, which is a close approximation to reality in many cases, then equation (1.19) holds with $B_f(t)$ replaced by $B(t)$, the *total* number of male and female zygotes. The only change is that $m(x, t)$ is replaced by $M(x, t)$ in equation (1.18). If the primary sex-ratio is a, then the numbers of females and males aged x at time $t(t \geq x)$ are given by

$$n(x, t) = aB(t - x)l(x, t) \tag{1.20b}$$

$$n^*(x, t) = (1 - a)B(t - x)l^*(x, t) \tag{1.20c}$$

where $l^*(x, t)$ is defined by an equation similar to (1.15), replacing $P(x - y, t - y)$ by $P^*(x - y, t - y)$.

1.2.2 *Continuous-time populations*

The models described above are exact only for populations which reproduce at discrete time-intervals, usually one year in length. There are, however, many species where there are no discrete seasons for reproduction, so that reproduction and other aspects of the population process are best thought of as taking place in continuous time. Human populations are the most familiar example, and the traditional apparatus of mathematical demography (Sharpe and Lotka, 1911; Lotka, 1925) was largely developed in an effort to understand and predict the growth of human populations. We first construct an exact description of the continuous-time model for the female population, on lines similar to the difference equation approach to the discrete age-class model described above. We then discuss how the continuous-time case can be adequately approximated by a suitable discrete-time formulation.

Exact formulation of the continuous-time model. Time t and age x are now taken to be continuous real variables; x occupies the closed interval $[0, d]$, where d is the upper limit to the reproductive ages. We can define a survival function $l(x, t)$ as the probability that a female survives from conception at time $t - x$ to age x at time t, and an *age-specific death-rate* $\mu(x, t)$ such that

$$\frac{\partial l(x, t)}{\partial x} + \frac{\partial l(x, t)}{\partial t} = -\mu(x, t)l(x, t) \tag{1.21}$$

With the condition $l(0, t) = 1$, this gives the relation

$$l(x, t) = \exp - \int_0^x \mu(y, t - y)\, dy \quad (t \geq x) \tag{1.22}$$

The probability of death in a small age-interval $[x, x + dx]$ is given by $\mu(x, t)\, dx$, and is clearly analogous to the age-specific probability of death $Q(x, t)$, defined for the discrete age-class model on p. 4. Equation (1.22) corresponds to equation (1.15).

We can similarly define the fecundity of a female aged x at time t as her rate of production of offspring $M(x, t)$, such that the number of offspring of both sexes which she is expected to produce in the small age-interval $[x, x + dx]$ is $M(x, t)dx$. More accurately, the number she is expected to produce in the interval $[x, x + x']$, given that she is alive at the end of the interval, is $\int_0^{x'} M(x + y, t + y)\, dy$. If the fraction of female zygotes among these zygotes is $a(x, t)$, her fecundity with respect to female offspring is defined by

$$m(x, t) = M(x, t)a(x, t) \tag{1.23}$$

The parallels with the discrete age-class formulae on pp. 3–4 are obvious. We can also define the *reproductive function* for the continuous-time model as $k(x, t) = l(x, t)m(x, t)$, by analogy with equation (1.16).

Finally, the state of the population at any time t can be described by the quantity $B_f(t)$; this is the rate of production of female zygotes by the population as a whole, such that the number of female offspring entering the population over a small time-interval $[t, t + dt]$ is $B_f(t)\, dt$. This will be referred to loosely as the *birth-rate* of the female population, although for genetic purposes we normally regard the population as being censused at the stage of zygote formation, before selection can act to alter its genetic make-up. Knowledge of $B_f(t)$ and the survival function $l(x, t)$ enables us to calculate the number of females in any age-interval. Let $n(x, t)$ be the density function for the number of females aged x at time t, such that the number aged between x and $x + dx$ at time t is $n(x, t)\, dx$. We clearly have (for $t \geqq x$)

$$n(x, t) = B_f(t - x)l(x, t) \tag{1.24a}$$

The number of females aged between x_1 and x_2 at time t is given by

$$n(x_1, x_2, t) = \int_{x_1}^{x_2} n(x, t)\, dx \tag{1.24b}$$

We are now in a position to relate $B_f(t)$ to the past state of the population; this can be achieved by means of an argument similar to that which leads to equations (1.17), (1.18) and (1.19). For times $t > d$, such that no females who were present at the initial time $t = 0$ are still reproductively active, we have

$$B_f(t) = \int_b^d B_f(t-x)k(x, t)\,\mathrm{d}x \qquad (1.25)$$

where b is the lower limit to the reproductive ages $(b > 0)$. For $t \le d$, it is necessary to take into account the contributions of females who were present at time $t = 0$. Let $n(0, x)\,\mathrm{d}x$ be the number of females aged between x and $x + \mathrm{d}x$ at this initial time. The contribution of the initial set of females to $B_f(t)$ is given by

$$g(t) = \int_t^d n(x-t, 0)\tilde{l}(x, t)m(x, t)\,\mathrm{d}x \qquad (1.26)$$

where

$$\tilde{l}(x, t) = \exp - \int_0^t \mu(x-y, t-y)\,\mathrm{d}y.$$

Combining this with equation (1.25) gives the continuous-time analogue of equation (1.19)

$$B_f(t) = g(t) + \int_0^t B_f(t-x)k(x, t)\,\mathrm{d}x \qquad (1.27)$$

where $g(t) = 0$ for $t > d$, and $k(x, t) = 0$ for $x > d$ and $x < b$. This is an example of an *integral equation*.

For most biological applications, it is reasonable to assume that $\mu(x, t)$ and $m(x, t)$ are continuous functions of x and t. It follows from this that $l(x, t)$ and $k(x, t)$ are also continuous in x and t. It is not necessary for $n(x, 0)$ to be continuous in x for equation (1.27) to be valid, but it must satisfy the standard mathematical conditions for integrability. If these conditions are met, then $g(t)$ and $B_f(t)$ are both continuous functions of t. (Continuity of $k(x, t)$ is, of course, not necessary for the validity of equation (1.27), and in fact equation (1.19) can be regarded as a special case of equation (1.27) if the theory of Lebesgue integration is applied. In an elementary treatment such as the present, it seems preferable, however, to treat the discrete-time and continuous-time models as distinct). Later on we shall see how equation (1.27) can be used to study the dynamics of the size and age-composition of populations (sections 1.3.1 and 1.4.1).

We may note that age-specific death-rates, survival functions and fecundities can be defined for males in exactly the same way as for females; as in the discrete age-class model, they will be distinguished by asterisks from their female counterparts. Furthermore, if the primary sex-ratio, a, can be treated as constant and independent of parental age, it is easy to show that equation (1.27) holds if $B_f(t)$ is replaced by the corresponding rate of production of male *and* female zygotes, $B(t)$, except that $m(x, t)$ in equation (1.26) must be replaced with $M(x, t)$. The parallel with the difference equation formulation of the discrete age-class model is, therefore, complete.

Discrete age-class approximations to continuous-time models. The discrete age-class model is obviously far more convenient for computational purposes than the continuous-time model. By dividing the continua of time t and age x into discrete intervals of arbitrary length, it is possible to approximate the continuous-time model to any desired degree of accuracy by a discrete age-class model, simply by making the widths of the intervals sufficiently small. For example, let the age interval $[0, d]$ be divided into m equal sub-intervals of width d/m. Consider a set of m distinct values of x, $\{x_i\}$ $(i = 1, 2, \ldots m)$, such that each x_i falls into one of the sub-intervals or on to a boundary between sub-intervals. (The positioning of the x_i with respect to the boundaries is arbitrary; in ecological applications, it is frequently the case that each x_i is taken as the mid-point of the ith sub-interval.) Equation (1.25) is then approximated by the expression

$$\tilde{B}_f(t) = \sum_{i=1}^{m} B_f(t - x_i)k(x_i, t)(d/m)$$

If time t is also divided into intervals of width d/m, we can approximate the continuous-time process described by $B_f(t)$ by the discrete-time process described by $\tilde{B}_f(j)$, where (for $j > m$) we have

$$\tilde{B}_f(j) = \sum_{i=1}^{m} \tilde{B}_f(j-i)\tilde{k}(i, j)$$

where $\tilde{k}(i, j) = k(x_i, t_j)(d/m)$.

Similar approximations can obviously be made to equation (1.27). The parameters of this discrete-time model are obtained directly from the continuous model. The probability of survival from age-class i to $i+1$ at the jth time-interval is given by $\tilde{P}(i, j) = l(x_{i+1}, t_{j+1})/l(x_i, t_j)$; the probability of survival from conception in the $i-j$th time-interval to age-class i in time-interval j is given by $\prod_{k=1}^{i} \tilde{P}(i-k, j-k)$; the fecundity of a female in age-class i in the jth time-interval is $m(x_i, t_j)(d/m)$, etc. With these definitions, and the appropriate change in notation, we can regard the matrix or difference equation formulations of section 1.2.1 as approximations to a continuous-time model, except that the unit of time is now the time-interval of arbitrary length d/m instead of one year. In practice, it is usually found that the continuous-time process is well approximated by the discrete model even if m is not very large. For example, in human demography it is conventional to take 5-year time-intervals as the basis for computation, thus dividing up the female reproductive and pre-reproductive life-span into 9 or 10 age-classes. The error resulting from this approximation seems to be negligible in comparison with other sources of error and artificialities of the model (Keyfitz, 1968, Chapter 8).

1.2.3 *The estimation of demographic parameters*

As we have seen above, the dynamic properties of a female population with discrete age-classes are determined by age-specific survival probabilities $P(x, t)$ (or, equivalently, by the survival function $l(x, t)$ and the age-specific fecundities, $m(x, t)$). Similar functions can be defined for a male population and, with appropriate modifications, for a continuous-time population. We may refer to these quantities collectively as the *demographic parameters* for the population in question. To apply our mathematical models to a real population, it is obviously necessary to possess estimates of the demographic parameters. Such estimates are usually referred to as the *vital statistics* or *vital rates* of the population. Obtaining satisfactory estimates presents considerable technical problems, whose detailed discussion is beyond the scope of this book. Interested readers should consult one of the standard works in this field, such as Harper (1977) for plants, Deevey (1947), Southwood (1966) or Caughley (1977) for animals, and Keyfitz (1968) or Henry (1976) for humans. Here we shall simply survey rapidly the basic ideas and methods involved in obtaining vital statistics, in order to provide some feel for the empirical basis of the models described in this chapter. For simplicity, the account is worded as if demographic parameters are completely independent of time and depend only on age and sex; it should be borne in mind that a set of vital statistics collected at any one time is frequently valid only for a limited period.

Direct methods of obtaining vital statistics. Two basic methods of directly collecting data, the 'horizontal' and the 'vertical' methods, may be distinguished. With the horizontal method, which is commonly used in the study of laboratory populations of animals and in the demography of historical human populations, the investigator follows the fate of a sample of individuals who were born at about the same time (a *cohort*), and records the times at which they die and the numbers of offspring which they produce at different ages. For example, if a laboratory strain of *Drosophila* is used as material, it is possible to follow the daily egg production of individual females throughout their life under standard culture conditions, and to record for each day the number of females surviving out of the set of newly eclosed females used to found the cohort. Separate experiments have to be conducted to determine the length of time from the egg stage to eclosion from the pupa, and the probability of survival from egg to eclosion. In this way it is possible to determine a set of values, $l(x)$, of probabilities of survival of a female from egg to day x; the fecundity $m(x)$ is expressed as one-half of the number of eggs a female alive at day x is expected to lay over one day (this assumes a sex-ratio of one-half). The table of values of $l(x)$ is known as the *life-table*, and the table of values of $m(x)$ as the *fecundity schedule*. Table 1.1 shows the life-tables and fecundity schedules for adult females of the species *Drosophila pseudoobscura* grown at 25 °C, both under optimal conditions and when yeast is omitted

Table 1.1. *Vital statistics for adult females of* Drosophila pseudoobscura

Age (x) (days)	Optimal conditions $l(x)$	$m(x)$	Yeast starvation $l(x)$	$m(x)$
0	0.60	0.0	0.46	0.0
4	0.60	23.6	0.46	0.3
8	0.60	20.2	0.44	1.1
12	0.60	25.8	0.44	1.2
16	0.60	25.1	0.43	0.5
20	0.59	16.4	0.41	0.8
24	0.53	16.6	0.39	0.7
28	0.46	17.0	0.38	0.6

Data of Anderson and Watanabe (1980) for females of karyotype *AR/AR*, reared at 25 °C. One hundred newly eclosed females were used to found each cohort. Measurements of larval and pupal viability showed that the probabilities of survival to time of eclosion ($l(0)$) were 0.60 and 0.46 for the optimal and starvation conditions respectively.

from the culture vials. In both cases, the mean time to eclosion is approximately 13 days, but the fecundity and survival of the flies are drastically reduced by starvation.

In the vertical method for obtaining vital statistics, the investigator uses census data on the same population at two successive times, e.g. one year apart. This is the standard method used in the demography of contemporary human populations, where very many data are available from national census figures. This type of data gives one the number of individuals of a given age-class that were alive at the start of the interval, and the number who died at some time during it. The number of offspring produced during the interval by individuals who belong to a given age-class at its start can similarly be estimated. There is extensive literature on the techniques for extracting vital statistics from census data of this sort, and of obtaining smoothed curves of $l(x)$ and $m(x)$ from censuses of continuous-time populations taken at discrete time-intervals (see Keyfitz, 1968). We shall take it for granted here that the technical problems involved can be satisfactorily overcome. Table 1.2 shows an example of a life-table and fecundity schedule for the U.S.

Table 1.2. *Vital statistics for U.S. females*
(*1964*)

Age-class (x) (5-year intervals)	$P(x)$	$l(x)$	$m(x)$
0	0.9779	1	–
1	0.9966	0.9746	0
2	0.9983	0.9730	0.0011
3	0.9979	0.9710	0.0898
4	0.9968	0.9679	0.3566
5	0.9960	0.9641	0.4868
6	0.9947	0.9590	0.3453
7	0.9923	0.9516	0.1875
8	0.9887	0.9408	0.0778
9	0.9830	0.9248	0.0178
10	–	0.9091	0.0010

These figures are taken from U.S. census data for 1964, as analysed by Keyfitz (1968, pp. 27–33). $P(0)$ is an estimate of the probability that a female child born at some time within a 5-year time interval will live to the end of the interval; $P(1)$ is an estimate of the probability that a female aged between 0 and 4 years will live for 5 years, $P(2)$ estimates the probability that a female aged between 5 and 9 years will live for 5 years, etc. Similarly, $m(2)$ estimates the number of daughters a female aged between 5 and 9 years is expected to bear over 5 years; $m(3)$ is the similar number for a female aged between 10 and 14 years, etc. $l(x)$ is obtained from equation (1.15).

female population of 1964. The data are grouped into discrete age-classes of 5 years in length, as is conventional in human demography.

Indirect methods of obtaining vital statistics. Ecologists interested in natural populations often have to resort to less direct methods for obtaining vital statistics. It is rarely possible, in dealing with a population which is living under natural conditions, to follow a cohort of individuals throughout their life, and to distinguish emigration of animals out of a study area from mortality. One method which is often used to get around these difficulties is to assume that the population is stationary in size and constant in age-structure. If this is the case, it is easy to see (equation (1.50a)) that the frequency of individuals in age-class x (in a discrete age-class population) is simply $l(x)/\sum l(x)$. If new-born individuals ($x = 0$) are included in the sample whose age-structure is measured, we can use the fact that $l(0)$ equals 1 to obtain the values of $l(x)$ from the frequencies of different ages in the sample. If new-born individuals cannot, for technical reasons, be sampled adequately, it is only possible to determine the values of $l(x)$ relative to the first age-class represented in the sample. An example of this method is provided by the work of Bulmer and Perrins (1973) on the great tit, who show that adult females (females aged 1 year and older) have an age-distribution which fits a life-table with an annual mortality of 0.52, independent of age. Males have a mortality of 0.44.

A related method, which also assumes a stationary population and constant age-structure, is to determine the distribution of age at death among a sample of dead individuals. For example, examination of the annual growth rings of the horns of mammals such as sheep enables one to determine the ages at death of a collection of skulls (Deevey, 1947; Caughley, 1966, 1977). If the population is stationary, it is easily seen that the frequency of individuals who die while in age-class x is $l(x) - l(x + 1)$, which enables one to determine the values of $l(x)$ relative to the value for the first age-class which can be reliably aged. A refinement of this method, which removes the necessity for assuming a stationary size for the population as a whole, is to mark individuals at birth or some convenient age, and then to attempt to recover as many marked corpses as possible. In this way, an estimate of the distribution of age at death can be obtained for a cohort. This method has been widely applied to the study of bird populations, owing to the ease of identifying individuals by ringing (Lack, 1954, 1966).

With all these indirect methods, supplementary observations must be made to determine the fecundity schedule, and also to estimate survival probabilities for ages which cannot be included in the basic sample material. Quite apart from the errors resulting from the fact that the assumptions of the methods are frequently violated in practice, considerable sampling errors are usually attached to the vital statistics obtained for natural populations, due to the difficulty of obtaining large samples. Relatively few really reliable sets of vital statistics are available for natural populations.

Enough data have been obtained, however, to give some insight into the types of life-histories adopted by different species, in terms of the way in which the vital statistics depend on age. We shall briefly consider some general features of the main types of life-history below.

Types of life-history.   There are two fundamentally different types of life-history, with respect to the pattern of reproduction as a function of age. These were named *semelparous* and *iteroparous* life-histories by Cole (1954). A species is said to have a semelparous life-history when reproduction is confined to a single age-class, after a more or less prolonged juvenile stage, and is followed by death. The most familiar type of semelparous life-history is that exhibited by annual species of plants or insects, where individuals reproduce in the breeding season following the year of their birth. There are also semelparous species in which reproduction takes place several years after conception. The pink salmon (*Oncorhynchus gorbuscha*), for example, breeds only at 2 years of age (Aspinwall, 1974). More extreme postponement of reproduction is exhibited by such organisms as the periodic cicada (*Magicicada*) of the eastern U.S.A., which reproduces and dies at 17 years (Lloyd and Dybas, 1966*a*, *b*; Bulmer, 1977). Semelparous species with long postponement of reproduction are also known in plants, e.g. the century plant *Agave americana*. There is thus considerable diversity among semelparous species with respect to the age of reproduction, although annual species are the commonest.

There is a wide range of life-histories among iteroparous species. They may reproduce in successive discrete breeding seasons (usually annual) as in many perennial plants and vertebrates; in other species, individuals produce successive litters at short intervals, without any synchronisation into breeding seasons (but usually with a suspension of reproduction in the winter, in temperate species); females may lay eggs continuously throughout their reproductive life, as in many tropical insects. There is considerable diversity even among quite closely related species with respect to such parameters as the age of first reproduction, man being an example of an iteroparous species which has postponed reproduction considerably, in comparison with his closest relatives among the great apes.

The relationship of fecundity with age is very variable among iteroparous species, and several different types of life-history can be distinguished. A pattern which is characteristic of many small birds and mammals is annual breeding, with individuals reproducing for the first time in the breeding season following the year of birth. It is often found that $M(x)$ (i.e. clutch size in birds or litter size in mammals) is more or less constant at all ages, with the possible exception of the first breeding season, where it is frequently smaller than later on. Similarly, the probability of survival over one year is generally much lower for juveniles than adults, whose annual survival rate is approximately independent of age. This type of pattern is exemplified by the great tit discussed above. A complete set of vital statistics for a natural population of the grey squirrel (*Sciurus carolinensis*), which illustrates this type of life-history for a mammal, is shown in Table 1.3. In the larger species of birds and

Table 1.3. *Vital statistics for female grey squirrels in North Carolina*

Age-class (x) (years)	$P(x)$	$l(x)$	$m(x)$
0	0.253	1	–
1	0.458	0.253	1.28
2	0.767	0.116	2.28
3	0.652	0.089	2.28
4	0.672	0.058	2.28
5	0.641	0.039	2.28
6	0.880	0.025	2.28
7	–	0.022	2.28

Data of Barkalow, Hamilton and Soots (1970). Annual survival was estimated by following marked individuals from the nestling stage; fecundity was determined by measurements of litter size (litters from females breeding in their first year of life have been ignored). Note that the survival probabilities for later ages are based on small samples.

mammals, the age of first reproduction is frequently more than one year, and they generally have a considerably higher survival rate. In fact, individuals may live so long that evidence for *senescence* may be observed, in the form of declining survival and fecundity with advancing age (Caughley, 1966). The human data of Table 1.2 illustrate an extreme type of life-history of this sort, where a peak level of fecundity is achieved relatively quickly after the start of reproduction, followed by a slow decline. Mortality increases at an accelerated rate after a minimum around the age of first reproduction. The *Drosophila* data shown in Table 1.1 are similar to this in general pattern. Both the human and *Drosophila* vital statistics are for species which under natural conditions would have much higher mortality rates, that would tend to conceal the senescent increase in mortality with age. Another type of reproductive pattern is shown by iteroparous cold-blooded vertebrates and many perennial herbs or woody plants (Harper and White, 1974), which continue to grow in size throughout reproductive life, unlike insects or female higher vertebrates. Since egg or seed number in these species is highly correlated with size, continued growth results in a positive correlation between age and fecundity. This is seen in many fishes, amphibia and reptiles (e.g. Hodder, 1963; Tinkle and Hadley, 1973).

This is by no means an exhaustive classification of life-history types, but is intended to give the reader an idea of the great diversity of possible patterns of reproduction and survival in relation to age. Some evolutionary forces which may have been important in moulding life-history patterns will be considered in Chapter 5. We now pass on to further mathematical development of the population models, using as illustrative material some of the sets of vital statistics presented in this section.

1.3   Time-independent and density-independent demographic parameters

So far we have been concerned only with the construction of a mathematical framework for describing the population process. A solution to the population process, in the sense of specifying its size and age-composition at some arbitrary time (given its initial state), can usually be obtained only when the age-specific survival probabilities are constant over time and independent of population size. This case will be considered in some detail in this section. Its importance does not lie in the fact that natural populations frequently satisfy these conditions; clearly they do not in general. Rather, many of the basic properties of more complex types of age-structured populations can be understood by reference to this case. We shall give most attention to the female population process described by equations (1.5), (1.19) and (1.27); the male population and the case of plants with seed dormancy are also considered briefly (section 1.3.3). Since it is assumed that the demographic parameters are constant over time, the argument t will be omitted from $P(x, t)$, $l(x, t)$, $m(x, t)$ etc. We consider in turn the matrix, difference equation and integral equation approaches, before developing some general results which are valid for all three methods.

1.3.1 *The dynamics of the population process for the female population*

The Leslie matrix solution. Consider first of all the process described by equations (1.5) and (1.6). The elements of the Leslie matrix L are now independent of time; it follows from equation (1.5) that we have

$$n(t) = n(0)L^t \tag{1.28}$$

If L has d linearly independent *eigenvectors*†, this equation can be written in terms of the *spectral expansion* of L^t

$$n(0)L^t = n(0) \sum_{i=1}^{d} \lambda_i^t V_i^T U_i \tag{1.29}$$

where λ_i is the ith *eigenvalue* of L, U_i is the corresponding *row eigenvector*, and V_i^T the corresponding *column eigenvector*, normalised such that $U_i V_i^T = 1$. (Readers unfamiliar with the results of matrix algebra employed here should consult a standard text on the subject; alternatively, Jacquard (1974, Appendix 2) provides a short summary.)

The eigenvalues $\lambda_i (i = 1, 2, \ldots d)$ are the roots of the *characteristic equation*

$$|L - \lambda I| = 0 \tag{1.30}$$

where I is the $d \times d$ unit matrix.

---

† A sufficient, but not necessary, condition for this is that L possesses d distinct eigenvalues.

The eigenvectors U_i and V_i^{T} are defined by the equations

$$U_i L = \lambda_i U_i \qquad (1.31a)$$

$$L V_i^{\mathrm{T}} = \lambda_i V_i^{\mathrm{T}} \qquad (1.31b)$$

If equation (1.30) is expanded, and the definitions of equations (1.3) and (1.15) are used, we obtain the following simple form for the characteristic equation†

$$\sum_{x=b}^{d} \lambda^{-x} k(x) = 1 \qquad (1.32)$$

(This equation can be derived in a straightforward way using the difference equation method, as we shall see on pp. 28–9.)

Equation (1.32) is sometimes called the *Euler equation*, owing to the derivation of a special form of it by Euler (1760). It is also frequently referred to as the *Lotka equation*, after A. J. Lotka (1925) who derived its continuous-time analogue and applied it to demographic problems.

The asymptotic solution. For large t, equation (1.29) is dominated by the term contributed by the *leading eigenvalue*, λ_1, say; this is the eigenvalue with greatest modulus (if λ_1 is complex, the term contributed by its complex conjugate must also be taken into account). In the case when the leading eigenvalue is real, equation (1.29) approaches, with increasing t, the expression

$$n(t) \sim n(0) \lambda_1^{t} V_1^{\mathrm{T}} U_1 \qquad (1.33)$$

By considering the corresponding expression for $n(t-1)$, it is seen that λ_1 gives the *asymptotic growth-rate* of the population; it is a quantity of fundamental importance in demographic theory. It can be shown from standard matrix theory (Pollard, 1973, Chapter 4) that there is a unique, real positive root of equation (1.32) which corresponds to λ_1, unless the age-specific net fecundities $f(x)$ are *periodic*. A periodic $f(x)$ function is defined here as one which is

---

† Many authors index the age-classes in the Leslie matrix 0, 1, 2, . . . instead of 1, 2, 3 . . . as here. An age-class x in such notation corresponds to age-class $x+1$ in the system used here, so that equation (1.32) takes a slightly different form, with an exponent of $\lambda^{-(x+1)}$ instead of λ^{-x}. Some texts incorrectly use an exponent of λ^{-x}.

non-zero only for x values which are a multiple of some fixed number other than one (e.g. reproduction occurring at ages 2, 4, 6, 8, etc.). The only periodic case likely to be encountered in biological applications is the semelparous life-history discussed above (p. 21). For iteroparous species, it is safe to assume that λ_1 is real and positive. These results can also be derived by the following elementary method, directly from the properties of the characteristic equation (1.32), using an approach due originally to Haldane (1927a).

Let λ_j be a root of equation (1.32), real or complex, and write

$$\lambda_j = e^{\phi_j + i\omega_j} = e^{\phi_j}(\cos \omega_j + i \sin \omega_j)$$

When $\omega_j = m\pi \, (m = 0, 1, 2, \ldots)$, λ_j is real; when m is zero or even, $\lambda_j > 0$; when m is odd, $\lambda_j < 0$. Separating real and complex parts of equation (1.32) with $\lambda = \lambda_j$, we obtain

$$\sum_x e^{-\phi_j x} \cos(-\omega_j x)k(x) = 1$$

$$\sum_x e^{-\phi_j x} \sin(-\omega_j x)k(x) = 0$$

Consider a real and positive λ_j, for which we simply have $\sum e^{-\phi_j x}k(x) = 1$. But $\sum e^{-\phi x}k(x)$ is easily seen to be a strictly decreasing function of ϕ, when ϕ is an arbitrary real variable. This implies that there is only one such real and positive λ_j. If $f(x)$ and hence $k(x)$ are periodic, it is always possible to choose $\omega = 2\pi/b$ such that the values of ωx for the reproductive age-classes are even multiples of π, since the only non-zero $k(x)$ values are those for which $x = b, 2b, 3b$ etc. For such an ω, we have $\cos(-\omega x) = 1$ and $\sin(-\omega x) = 0$ for $x = b, 2b$, etc., so that equation (1.32) is satisfied with $\lambda = e^{\phi_j + i\omega}$. Only if ω is itself an even multiple of π is this root equal to λ_j; this occurs if and only if $b = 1$, which is excluded from the category of periodic $k(x)$. This proves that in the periodic case it is always possible to find a negative real root or complex root of equation (1.32) which is equal in modulus to the positive real root λ_j.

If $k(x)$ is not periodic, it is not possible to choose an ω such that $\cos(-\omega x) = 1$ for all reproductive ages. Suppose that there is a complex root λ_k such that $\phi_k \geqq \phi_j$, i.e. the modulus of λ_k is greater than or equal to that of λ_j. Then, since $|\cos(-\omega_k x)| < 1$ for at least some values of x, we have

$$\sum_x e^{-\phi_k x} \cos(-\omega_k x) k(x) < \sum_x e^{-\phi_k x} k(x) \leqq 1$$

since $e^{-\phi_k x} \leqq e^{-\phi_j x}$ for all x. Hence λ_k cannot be a root of equation (1.32). λ_j is therefore the root of largest modulus, giving the conclusion that there is a real, positive leading eigenvalue λ_1 in the non-periodic case.

Further consideration of the periodic case will be deferred until p. 29. We can therefore assume that L possesses a leading eigenvalue λ_1, which is given by the positive real root of equation (1.32). For t taken sufficiently large, equation (1.33) is the solution of the population process.

Since, by equation (1.33), each age-class asymptotically grows in size at the geometric rate λ_1 every time-interval, it follows that the age-structure of the population, in terms of the relative numbers of individuals in the different age-classes, is constant. The population is said to have a *stable age-distribution* when this state has effectively been reached. The concept of stable age-distribution plays an important role in demographic theory, and will be discussed in detail in section 1.3.2.

Before turning to the solution of the population process by the difference equation method, we may note that equation (1.29) assumes that the eigenvectors of L are linearly independent. This is usually found to be the case in practical applications (Keyfitz, 1968, p. 51); if it is not so, a biologically trivial perturbation to some of the elements of L will usually suffice to make it true. It is useful to note that, even if this condition is violated, there is always a positive, real leading eigenvalue in the non-periodic case, and that the asymptotic result of equation (1.33) is valid.

The difference equation solution.   In this section, we show how the population process can be solved by means of the difference equation formulation introduced on pp. 10–12. If there is a time-independent reproductive function $k(x)$, equation (1.19) becomes simply

$$B_f(t) = g(t) + \sum_{x=1}^{t} B_f(t-x) k(x) \tag{1.34}$$

This is an example of a *renewal equation*, a class of linear difference equations whose properties have been much studied by mathematicians interested in probability theory and related areas. In the present case, when $k(x) = 0$ for $x > d$ and $g(t) = 0$ for $t \geqq d$, equation (1.34) can be approached by the method of *generating functions*, which is outlined in Appendix 1. Taking generating functions on both sides of this equation, and defining $k(0)$ as zero, we obtain

$$\bar{B}_f(s) = \bar{g}(s) + \bar{B}_f(s)\bar{k}(s) \tag{1.35}$$

so that

$$\bar{B}_f(s) = \frac{\bar{g}(s)}{1 - \bar{k}(s)} \tag{1.36}$$

This equation is of the same form as equation (A1.3) of Appendix 1; it follows that, providing that the zeros of $1 - \bar{k}(s)$ are all distinct, we have

$$B_f(t) = \sum_{j=1}^{d} C_j s_j^{-(t+1)} \tag{1.37}$$

where the C_j are given by equation (A1.6) of Appendix 1. Asymptotically we have

$$B_f(t) \sim C_1 s_1^{-(t+1)} \tag{1.38}$$

where s_1 is the zero of $1 - \bar{k}(s)$ with smallest modulus. But $\bar{k}(s) = \sum s^x k(x)$, so that s^{-1} can be identified with λ of equation (1.32). λ_1 is therefore equal to s_1^{-1}. Evaluating the differential coefficient of $1 - \bar{k}(s)$ with respect to s at s_1, and using equation (A1.6) of Appendix 1 together with equation (1.38), gives the asymptotic result

$$B_f(t) \sim \lambda_1^t \sum_{x=0}^{d-1} \lambda_1^{-x} g(x) / \sum_{x=1}^{d} x\lambda_1^{-x} k(x) \tag{1.39}$$

As with the matrix approach, this asymptotic solution is valid even if $1 - \bar{k}(s)$ has repeated zeros, provided that $k(x)$ is non-periodic (see Appendix 1). The numbers of individuals in any age-class at a given time t can be obtained from equation (1.39) by using equation (1.20). Equations (1.33) and (1.39) must therefore correspond; the nature of this correspondence will be discussed below (pp. 37–9).

It is useful to note that equation (1.32), from which λ_1 is obtained, can be derived directly from equation (1.34). If the population has attained its stable age-distribution and is growing at rate λ_1, we have $B_f(t) = C\lambda_1^t$, where C is a constant of proportionality. Substituting from this into equation (1.34) for $t \geqq d$, we find

$$C\lambda_1^t = C \sum_x \lambda_1^{(t-x)} k(x)$$

so that, cancelling $C\lambda_1^t$ from both sides, we see that λ_1 must satisfy equation (1.32).

The difference equation approach to the population process has the slight advantage over the matrix method that it can be applied to situations in which reproduction continues indefinitely, without a finite upper limit d to the fertile age-classes. Techniques which are more advanced than those employed here can be used to show that, provided $k(x)$ is non-periodic and $\sum_{t=0}^{\infty} g(t)$ is finite, equation (1.34) has the asymptotic solution (1.39). The interested reader should consult Feller (1968, Chapter 13) for a full discussion of this result, the *renewal theorem*.

The periodic case. As mentioned earlier, the only periodic case of much biological interest is the semelparous form of life history. Most semelparous species effectively consist of a number of non-interbreeding populations (phases) represented in different years, the individuals born in year $t - b$ becoming the parents of individuals born in year t. The number of female births in year t is therefore connected with those in year $t - b$ by the relation

$$B_f(t) = B_f(t - b)k(b) \quad (t = b, b + 1, \dots) \tag{1.40}$$

For $t \geqq b$, the number of births corresponding to a given phase of the population increases at a geometric rate $k(b)$, the numbers for $t < b$ being given by the initial conditions. In certain cases, of which the most famous is probably the 17-year cicada, *Magicicada* (Lloyd and Dybas, 1966*a*, *b*; Bulmer, 1977), reproduction has been synchronised so that a breeding population appears only every b years ($b = 17$ for *Magicicada*). The mechanism of this synchronisation will be discussed in Chapter 5 (pp. 229–31).

The continuous-time model. The continuous-time model with a time-independent reproductive function yields the continuous version of the renewal equation as a special case of equation (1.27):

$$B_f(t) = g(t) + \int_0^t B_f(t-x)k(x)\,\mathrm{d}x \qquad (1.41)$$

It is easily seen that (for $t > d$, so that $g(t) = 0$) this equation is satisfied by substituting $B_f(t) = C\,\mathrm{e}^{zt}$, where z satisfies the characteristic equation analogous to equation (1.32)

$$\int_b^d \mathrm{e}^{-zx}k(x)\,\mathrm{d}x = 1 \qquad (1.42)$$

Using the argument applied above to equation (1.32), it can be shown that equation (1.42) has a single real root, z_1 say, which exceeds the real parts of the complex roots; it is sufficient for this to be true that $k(x)$ be continuous, non-negative, and take some positive values between b and d. This condition is normally satisfied in biological applications. By analogy with the results for the difference equation approach to the discrete-time case, one might expect equation (1.41) to have a general solution of the form

$$B_f(t) = \sum_{i=1}^{\infty} C_i\,\mathrm{e}^{z_i t} \qquad (1.43)$$

where the z_i are the roots of equation (1.42), assumed to be distinct, and the C_i are constants given by the initial conditions. This is the form of the solution proposed by Lotka and his co-workers (Lotka, 1925). If such a solution in fact exists, then (using the fact that the real root z_1 is larger than the real parts of all the complex roots) it follows that equation (1.43) becomes asymptotically

$$B_f(t) \sim C_1\,\mathrm{e}^{z_1 t} \qquad (1.44a)$$

so that the logarithm of $B_f(t)$ eventually grows at constant rate z_1. It follows from equations (1.24) that the same is true for the number of individuals in any age-group, so that the population asymptotically achieves a state of stable age-distribution, just as in the discrete age-class case.

There are some mathematical problems in rigorously justifying the solution represented by equation (1.43). These are discussed by

Feller (1941) and Lopez (1961). It seems in practice to provide an adequate answer in most cases of biological interest and (according to the results of Lopez) is valid if $k(x)$ is continuous and positive for values of x between b and d, assuming that all the zeros of equation (1.42) are distinct. The asymptotic expression (1.44a) is known, in fact, to be valid under very wide mathematical conditions, which subsume those which are sufficient for equation (1.42) to have a single real root, larger than the real parts of the complex roots (Feller, 1966, Chapter 11). The constant C_1 has the value

$$C_1 = \int_0^d e^{-z_1 x} g(x)\, dx \Big/ \int_b^d x\, e^{-z_1 x} k(x)\, dx \qquad (1.44b)$$

which shows the analogy between equations (1.39) and (1.44a).

1.3.2 *Populations with stable age-distributions*

As we have seen, both discrete age-class and continuous-time populations converge with increasing time to a state in which the age-structure and rate of population growth are constant. A population which has attained this state of stable age-distribution tends to return to it if it suffers a perturbation. The age-structure and growth-rate are independent of the initial state of the population, and are determined only by the values of the demographic parameters. This is the so-called *strong ergodicity* property of age-structured populations. We now discuss in some detail the properties of populations in stable age-distribution.

The intrinsic rate of increase, generation time and expectation of life. As we have seen, a discrete age-class population in stable age-distribution grows at the geometric rate λ_1, given as the positive real root of equation (1.32); a continuous-time population grows at the logarithmic rate z_1, given by the analogous equation (1.42). In order to unify the discrete-time and continuous-time cases, it is often convenient to work with the natural logarithm of λ_1 rather than λ_1. This is designated by the symbol r, and can be identified with z_1 of the continuous-time case. It has been called the *intrinsic rate of natural increase*, (Lotka, 1925) or the *Malthusian parameter* (Fisher, 1930). Here, r will usually be referred to simply as the *intrinsic rate of increase*; it measures the asymptotic logarithmic rate of growth in both the discrete-time and continuous-time models. In the former case, r can be determined numerically (for example, by Newton–Raphson iteration), using the fact that, by equation (1.32), we have

$$\sum e^{-rx} k(x) = 1 \tag{1.45}$$

This equation is analogous to equation (1.42) for the continuous-time model, which can similarly be solved by a combination of numerical integration and Newton–Raphson iteration. In most cases, equation (1.45) is too complex for r to be evaluated explicitly. If r is sufficiently small that terms of order r^2 can be neglected, an approximate expression for r can be obtained by noting that a Taylor expansion of equation (1.45) gives

$$1 = \sum e^{-rx} k(x) = \sum k(x) - r \sum x k(x) + \cdots$$

so that

$$r \cong \frac{\sum k(x) - 1}{\sum x k(x)} \tag{1.46}$$

The quantity $\sum k(x)$ in this expression is the net expectation of female offspring to a female zygote, and is called the *net reproduction rate*. It will be denoted by the symbol R. It follows from the properties of equation (1.45) that R must exceed unity for r to be positive. If $R = 1$, the stable age-distribution corresponds to a population which is stationary in size ($r = 0$); if $R < 1$, then the stable age-distribution corresponds to a decreasing population ($r < 0$).

The quantity $\sum x k(x)$ in the denominator of equation (1.46) is the mean age at reproduction for a female in a stationary population, and may be used as one measure of 'generation time'. The above estimate of r is clearly unreliable if $r \sum x k(x)$ is more than a few per cent i.e. if the per 'generation' rate of population increase is high. For slowly growing populations, it provides an adequate approximation. For example, with the grey squirrel data quoted in section 1.2.3, we find that $R = 1.119$ and $\sum x k(x) = 3.332$, giving an approximate value for r of 0.036. The value obtained by Newton–Raphson iteration of equation (1.45) is 0.041, in reasonably good agreement.

Since the concept of generation time is a rather arbitrary one in the context of age-structured populations, several alternative measures have been proposed. When r is high, a more appropriate measure than $\sum xk(x)$ is the quantity

$$T_0 = \sum xk(x)/\sum k(x) \qquad (1.47a)$$

This is the mean age at reproduction of a cohort of females, and is independent of any assumptions about the rate of population growth. Another definition of generation time is the value T_1 which satisfies the equation

$$e^{-rT_1}R = 1$$

i.e.

$$T_1 = (\ln R)/r \qquad (1.47b)$$

Finally, another definition of generation time is the mean age of the mothers of a set of new-born individuals in a population with a stable age-distribution. Using equations (1.39) and (1.20) it is easily seen that this mean age is given by the expression

$$T = \sum x\, e^{-rx}k(x) \qquad (1.47c)$$

which appears in the denominator of equation (1.39). T is in fact the modulus of the derivative of equation (1.45) with respect to r. As will be seen shortly, T arises naturally in a number of contexts, and is therefore to be preferred to other measures of generation time. Provided r is not too big, all these measures are approximately equal. Using the grey squirrel data, for example, we find $T_0 = 2.794$, $T_1 = 2.273$, and $T = 2.676$.

Generation time should not be confused with *expectation of life*, E, which is the expected value of the time of death of a new-born individual. If an individual who dies between ages $x - 1$ and x is classified as dying in the xth interval of life, and noting that $l(0)$ is defined as equal to one, we have

$$E = \sum_{x=1}^{d} x[l(x-1) - l(x)] = \sum_{x=0}^{d} l(x) \qquad (1.48)$$

Since this measure takes no account of reproduction, it may differ considerably in value from the generation times. For example, in the grey squirrel case $E = 1.602$. In species with high juvenile losses, E very largely reflects the fact that most individuals die in the first age-class. This measure therefore gives a biased picture of the life-span. The bias can be corrected to some extent by defining a *conditional expectation of life* for individuals who have survived to age x, E_x, by the relation

$$E_x = \sum_{y=x+1}^{d} (y-x)[l(y-1) - l(y)]/l(x) \qquad (x = 1, 2, \ldots)$$

$$= \sum_{y=x}^{d} l(y)/l(x) \tag{1.49}$$

In the squirrel case, for example, we find $E_1 = 2.379$, showing a considerable increase in expectation of life over E.

All these quantities can be defined for the continuous-time model simply by replacing summation over age-classes by integration over age.

The form of the stable age-distribution. The form of the stable age-distribution of a discrete age-class population is defined by the frequencies of individuals falling into the different age-classes. Using equations (1.39) and (1.20), it is easy to see that the frequency of individuals in age-class x, $\phi(x)$, is given by the expression

$$\phi(x) = e^{-rx}l(x) / \sum_{x=1}^{d} e^{-rx}l(x) \tag{1.50a}$$

Using equation (1.33) and the definition of U_1, the row eigenvector of L associated with the leading eigenvalue λ_1 (equation (1.31a)), it is evident that the components of U_1, $U_1(1)$, $U_1(2)$, ... $U_1(d)$, are proportional to $\phi(1)$, $\phi(2)$, ... $\phi(d)$.

With the continuous-time model, we can similarly define a probability density function $\phi(x)$, which gives the frequency of individuals between ages x and $x + dx$ as $\phi(x)\,dx$. This has the form

$$\phi(x) = e^{-rx}l(x) / \int_0^d e^{-rx}l(x)\,dx \tag{1.50b}$$

The shape of the stable age-distribution is, from equations (1.50), obviously determined both by the intrinsic rate of increase and by the shape of the $l(x)$ function. High fecundity, resulting in high values of r, leads to a high weight being placed on early ages, so that the age-structure of the population is dominated by younger individuals. Similarly, if mortality rates are high throughout life, $l(x)$ will fall off rapidly with age, giving a similar effect. The form of the stable age-distribution is determined solely by the form of the $l(x)$ function only in stationary populations, where $r = 0$. Provided $r \gtreqless 0$, $\phi(x)$ is always a decreasing function of x.

It is interesting to note that there is no difference in the form of the stable age-distribution between two populations whose demographic parameters differ only in a component of survival which is the same for each age. Thus, if one population has the set of age-specific survival probabilities $P(x)(x = 1, 2, \ldots)$ and the other the set $P'(x) = KP(x)$ where K is independent of age, we have $l'(x) = K^x P(x)$. Substituting from this into equation (1.32), it is easily seen that the value of λ_1 for the second population is equal to that for the first multiplied by K. Using equation (1.50a), it is seen that $\phi(x)$ is the same for both populations. This type of difference in the life-table is what one would expect if the populations differed with respect to some 'accidental' source of mortality which is indiscriminate with respect to age. Coale (1957, 1972) has shown that many human populations have changed demographically over time largely in respect to mortality factors of this sort; they have therefore remained close to a single stable age-distribution.

Changes in fecundity have quite different effects on the age-distribution; since they alter the value of r, they must change $\phi(x)$ unless compensated for by changes in mortality. This suggests that changes in age-specific survival probabilities have a much smaller effect on the age-structure of a population than do changes in fecundity, unless they are highly selective with respect to age (e.g. if juvenile survival alone is changed).

Per capita birth-rates and death-rates. The *per capita birth-rates* and *death-rates* for a discrete age-class population are defined as the total numbers of new births and deaths, respectively, occurring over a single time-interval, divided by the total number of individuals alive in the time-interval in question. A population with a stable age-distribution must clearly have constant values for these quantities. If the new-born individuals are counted at the beginning of the time-interval *subsequent* to that in which they were born, we obtain, using equation (1.50a), the following expression for the per capita birth-rate, B, of a population in stable age-distribution

$$B = l(1) \sum_{x=b}^{d} e^{-rx} k(x) / \sum_{x=1}^{d} e^{-rx} l(x)$$

$$= l(1) / \sum_{x=1}^{d} e^{-rx} l(x) \tag{1.51a}$$

Similarly, the per capita death-rate for the stable population is given by

$$D = \sum_{x=1}^{d} e^{-rx} [l(x) - l(x+1)] / \sum_{x=1}^{d} e^{-rx} l(x) \tag{1.51b}$$

The geometric rate of population growth for the stable population is related to $B - D$; since each age-class increases in number at the same geometric rate λ_1, the total number of individuals in the population at time t, $n(t)$, is related to the number at time $t-1$ by the expression

$$n(t) - n(t-1) = (\lambda_1 - 1) n(t-1)$$

But from the definitions of B and D, we must also have

$$n(t) - n(t-1) = (B - D) n(t-1)$$

These equations show that $B - D$ is equal to $\lambda_1 - 1$; this can be shown in a different way by noting that equation (1.51b) gives us

$$D \sum_{x=1}^{d} \lambda_1^{-x} l(x) = \sum_{x=1}^{d} \lambda_1^{-x} l(x) - \lambda_1 \sum_{x=1}^{d} \lambda_1^{-x} l(x) + l(1)$$

which can be combined with equation (1.51a) to yield $B - D = \lambda_1 - 1$.

Similar quantities can be defined for continuous-time populations; by analogy with equations (1.51) we have

$$B = \int_0^d e^{-rx} k(x) \, dx \Big/ \int_0^d e^{-rx} l(x) \, dx \tag{1.52a}$$

$$D = \int_0^d e^{-rx} \mu(x) l(x) \, dx \Big/ \int_0^d e^{-rx} l(x) \, dx \tag{1.52b}$$

Integrating equation (1.52*b*) by parts and combining it with equation (1.52*a*) verifies that $r = B - D$, as one would expect from the definition of *r* as the logarithmic rate of increase.

A population which is not in stable age-distribution will not in general have constant per capita birth-rates and death-rates. It is a well-known fact of demography that direct estimates for *B* and *D* for a population which is in the process of approaching a stable age-distribution may be poor guides to their final values, owing to changes in age-structure which alter the weights attached to ages with different survival rates and fecundities (see Figures 1.1 and 1.2).

Reproductive value. Fisher (1930) introduced the concept of *reproductive value* as a measure of the relative extents to which individuals of different ages, in a population in stable age-distribution, contribute to the ancestry of the future population. In a discrete age-class population, reproductive value is closely connected with the column eigenvector V_1^T defined by equation (1.31*b*), as was first pointed out by Leslie (1948). This can be seen as follows. From equations (1.5), (1.6) and (1.31*b*), we obtain the following relation, which enables us to determine *relative* values of the components $V_1(1)$, $V_1(2)$, ... $V_1(d)$ of V_1^T

$$\lambda_1 V_1(x) = f(x) V_1(1) + P(x) V_1(x+1) \quad (x = 1, 2, \ldots d) \qquad (1.53)$$

(Note that $V_1(d+1)$ must be defined as zero).

It is easily verified that equation (1.53) is satisfied if we write $V_1(x) = Cv(x)$, where *C* is a constant of proportionality and $v(x)$ is the reproductive value of age *x*, defined by equation (1.54*a*)†

---

† If the notation referred to at the foot of p. 25 is used (with the first age-class given the index zero instead of one), it is preferable to define $v(x)$ relative to a value of unity for an individual who has survived one time-interval since conception, and is thus in age-class 0. This yields this formula for reproductive value:

$$v(x) = \left[e^{rx} \sum_{y=x}^{d-1} e^{-r(y+1)} k(y) \right] / l(x)$$

where age-class *x* is equivalent to age-class $x+1$ in equation (1.54*a*). The reproductive value for a given age in this system is equal to $e^{-r}P(0)$ times its value in the system used in the text.

$$v(x) = \frac{e^{rx}}{l(x)} \sum_{y=x}^{d} e^{-ry} k(y) \quad (x = 1, 2, \ldots d) \tag{1.54a}$$

Note that from equations (1.32) and (1.15) we have

$$v(1) = e^r / P(0) \tag{1.54b}$$

This is the reproductive value of an individual who has survived for one time-interval after conception. We can use equation (1.54a) to define the reproductive value of a newly produced female zygote (noting that $l(0) = 1$) as

$$v(0) = 1 \tag{1.54c}$$

The reproductive value of a female aged x on this scale can be thought of as measuring her expected future contribution of zygotes to the population, counting both the offspring contributed at age x and those she is expected to contribute if she survives to later ages. Since the population increases in number by a factor of e^r every time-interval, the value of her contribution at a future age $x + y$ is discounted by e^{-ry}. Reproductive value is thus measured relative to a value of unity for a newly-produced female zygote.

Since only the relative values of the components of V_1^T are determined by equation (1.31b), the above choice of scale for reproductive value is arbitrary, but turns out to be convenient for genetic purposes (see Chapters 3 and 4). An alternative scale of measurement, which arises naturally in a demographic context, has been proposed by Goodman (1967). Consider the asymptotic expression for the number of individuals in age-class y at time t, given by equation (1.33)

$$n(y, t) \sim e^{rt} U_1(y) \sum_{x=1}^{d} n(x, 0) V_1(x) \tag{1.55a}$$

where $U_1(y)$ and $V_1(x)$ are yth and xth components of the vectors U_1 and V_1^T, when normalised such that $U_1 V_1^T = 1$ (p. 24). We know from p. 34 that $U_1(y)$ is proportional to $e^{-ry} l(y)$; for simplicity, let the constant of proportionality be unity. We know from the above that $V_1(x) = Cv(x)$, where C is given by the normalisation requirement. Using this, together with equations (1.54), gives

$$1 = C \sum_{x=1}^{d} e^{-rx} l(x) v(x) = C \sum_{x=1}^{d} \sum_{y=x}^{d} e^{-ry} k(y)$$

i.e.

$$1 = C \sum_{x=b}^{d} x \, e^{-rx} k(x)$$

We therefore have

$$V_1(x) = v(x)/T \tag{1.56}$$

where T is the generation time defined by equation (1.47c). Equation (1.55a) can thus be rewritten as

$$n(y, t) \sim e^{r(t-y)} l(y) \left(\sum_{x=1}^{d} n(x, 0) v(x)/T \right) \tag{1.55b}$$

This may be compared with the difference equation solution of equation (1.39). It is easy to show from the definition of g in equation (1.18) that we have

$$\sum_{x=1}^{d} n(x, 0) v(x) = \sum_{x=0}^{d-1} e^{-rx} g(x)$$

Furthermore, from equation (1.20) we have $n(y, t) = B_f(t - y) l(y)$. Substituting these expressions into equation (1.39), we recover equation (1.55b), thus unifying the matrix and difference equation solutions.

The normalisation of reproductive value by the reciprocal of the generation time in equation (1.56) gives the quantity which Goodman (1967) has called the *eventual reproductive value* of age x. It is a measure of the extent to which individuals aged x at time zero contribute to the ancestry of the population at some distant time t, when the population has approached stable age-distribution. It is thus conceptually somewhat different from Fisher's reproductive value, as developed above for the discrete age-class model, despite the fact that the two differ only by a constant of normalisation.

Reproductive value as defined by equations (1.54) or (1.56) does not have a simple monotonic relationship with age. If r is non-negative it follows from the fact that $l(x)$ is a decreasing function of

x that reproductive value for the juvenile age-classes is an increasing function of age. But, for the reproductive age-classes, it is evident that $\sum_{y=x}^{d} e^{-ry} k(y)$ decreases with increasing x, so that reproductive value may decrease with age for at least some reproductive ages. With sufficiently high r and no upper limit to the reproductive age-classes ($d = \infty$), examples can, however, be constructed in which reproductive value increases indefinitely with age (Hamilton, 1966). Table 1.4 shows the relationship with age of

Table 1.4. *Reproductive value in two populations*

Age-class (x)	U.S. humans (1964) $v(x)$	Grey squirrels $v(x)$
0	1	1
1	1.106	4.119
2	1.201	6.454
3	1.300	5.670
4	1.312	5.420
5	1.037	4.866
6	0.586	4.205
7	0.184	2.280
8	0.093	–
9	0.018	–
10	0.001	–

reproductive value for two of the sets of vital statistics discussed earlier in this chapter. The differences between the human and squirrel cases shown here can be understood qualitatively in terms of the differences in the ways in which their vital statistics depend on age (see Tables 1.2 and 1.3). The human population is subject to very low mortality, so that there is only a slight increase in reproductive value during pre-reproductive and early reproductive life, compared with the squirrel population. On the other hand, the fecundity of human females falls off rapidly from age-class 5 onwards, whereas squirrel fecundity is constant. Reproductive value therefore tends to decrease more sharply with advancing age for older human females than for squirrels. The significance of the behaviour of reproductive value as a function of age, in connection with evolutionary problems, will be discussed in Chapter 5 (pp. 215–16 and 265–7).

One final property of reproductive value is of theoretical interest. We can define the *total reproductive value*, $V(t)$, of an arbitrary population (which is not necessarily in stable age-distribution) at time t by the relation

$$V(t) = n(t) V_1^T \qquad (1.57)$$

Using equation (1.5), it can be seen that this is related to the population at time $t - 1$ by the expression

$$V(t) = n(t-1) L V_1^T$$

and using equation (1.31b), this gives

$$V(t) = \lambda_1 n(t-1) V_1 = \lambda_1 V(t-1) \qquad (1.58)$$

In other words, the total reproductive value of a population increases at a geometric rate $\lambda_1 = e^r$, regardless of whether or not it is in stable age-distribution.

The concept of reproductive value can equally well be defined for a continuous-time population, and in fact this is how Fisher (1930) originally developed it. By analogy with equation (1.54a) we can write

$$v(x) = \frac{e^{rx}}{l(x)} \int_x^d e^{-ry} k(y) \, dy \qquad (1.59)$$

Using the continuous-time solution to the population process given by equations (1.44), we can obtain the analogue of equation (1.55b)

$$n(y, t) \sim e^{r(t-y)} l(y) \times$$
$$\int_0^d e^{-rx} n(x, 0) v(x) dx \Big/ \int_b^d x \, e^{-rx} k(x) \, dx \qquad (1.60)$$

Similarly, if we define the total reproductive value of a population as $V(t) = \int_0^d n(x, t) v(x) \, dx$, it is possible to show that

$$\frac{dV}{dt} = rV \qquad (1.61)$$

independently of the age-structure of the population. This result was derived by Fisher (1930); a concise proof is given by Crow and Kimura (1970, pp. 20–2).

Convergence to the stable age-distribution. Obviously the properties of the stable age-distribution which we have just discussed are not of much interest unless real populations converge towards stability in a reasonably short time. The rate of convergence can be most directly studied using the difference equation solution given by equation (1.37), which can be rewritten as

$$e^{-rt}B_f(t-1) = C_1 + \sum_{j=2}^{d} C_j e^{(z_j-r)t} \qquad (1.62)$$

where the e^{z_j} (with $z_j = \phi_j + i\omega_j$) are the roots of equation (1.32) other than λ_1. Each complex root λ_j has a conjugate root, $e^{\phi_j - i\omega_j}$, whose C coefficient is the conjugate of C_j. Each pair of complex conjugate roots therefore contributes an oscillatory term of the form

$$e^{(\phi_j-r)t}[C'_j \cos(\omega_j t) + C''_j \sin(\omega_j t)]$$

to the right-hand side of equation (1.62). Similarly a real, negative root, λ_k say, contributes a term of form

$$C_k e^{(\phi_k-r)t} \cos(\pi t)$$

which oscillates between positive and negative values in successive time-intervals. The amplitudes of these oscillatory terms diminish at a rate determined by the values of $\phi_j - r$, which we have already shown to be negative (pp. 26–7). The constant term in equation (1.62) comes to predominate after a time that is governed by the magnitude of the difference between r and the largest of the ϕ_j, ϕ_2 say. During a period of time of order $1/(r-\phi_2)$, the smaller roots still contribute significantly, and generate waves in the value of $B_f(t)$, and hence in the numbers of individuals in other age-classes. Keyfitz (1968, Chapter 6; 1972) discusses the period of these waves and the rate at which they damp out, for the type of reproductive function characteristic of human populations. The complex part ω_2 of the largest complex root is approximately equal to $2\pi/T_0$, where T_0 is the measure of generation time defined by equation (1.47a). The periodicity of the waves in a population which is converging to its stable age-distribution is thus of the order of the 'generation time'. The modulus of the real part ϕ_2 is such that effective convergence to the stable age-distribution takes place within approximately five generations. This high speed of convergence seems to be fairly common, on the basis of the numerical examples which have been studied in the literature. Figures 1.1 and 1.2 show two examples, using the sets of vital statistics quoted earlier.

Figure 1.1. The approach to stable age-distribution for a population with the demographic parameters of Table 1.2 (U.S. 1964 females). (*a*) The histograms show the frequencies of individuals in age-classes 1, 3, 5, 7 and 9, at successive time-intervals of 5 years. (*b*) The graph shows the corresponding values of the population growth-rate, as measured by $n(1, t+1)/n(1, t)$. The initial population was such that each age-class was equally frequent.

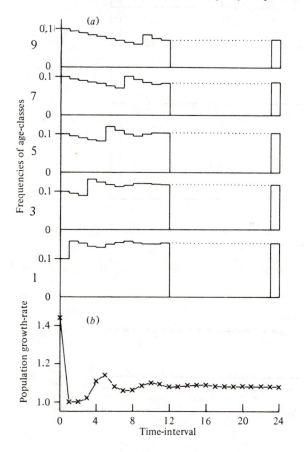

Figure 1.2. The approach to stable age-distribution for a population with the demographic parameters of Table 1.3 (N. Carolina female squirrels). (*a*) and (*b*) are as for Figure 1.1, except that time-intervals in this case are 1 year in length.

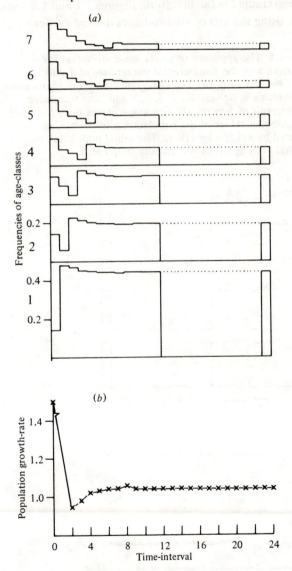

1.3.3   The dynamics of the population process for more complex types of population

The analysis given above is valid only for the female segment of the population, or (with appropriate notational changes) for a purely hermaphrodite population. We shall now consider briefly how it can be extended to the male segment of the population, and to the case of a species of plants with a distinct seed population (see pp. 8–10).

The male population. On the assumption of time-independent demographic parameters, and the model leading to equation (1.12), it is clear that the male population must approach an asymptotic state with a stable age-distribution at the same rate as the female population. This follows from the fact that, on the assumption of time-independent age-specific sex-ratios $a(x)$ (which is necessary for a time-independent $m(x)$ function), the number of male offspring produced in a time-interval is completely determined by the number of females of different ages. As the age-structure of the female population approaches constancy, therefore, the number of male births in a time-interval, $B_m(t)$, must approach the same rate of increase, λ_1, as the number of female births, $B_f(t)$. More formally, it is known from standard theory that the set of eigen-values of the population projection matrix in equation (1.12) (with time-independent demographic parameters) is identical with the set of eigenvalues of the sub-matrices L and B; it can easily be shown the latter are all zero, so that the leading eigenvalue of the matrix for the population of males and females is λ_1, giving the asymptotic rate of increase of both males and females. It follows that the proportion of males in a population which is effectively in stable age-distribution is given by an equation similar in form to equation (1.50a), with $l(x)$ replaced by $l^*(x) = \prod_{y=1}^{x} P^*(x-y)$. Furthermore, in a population in stable age-distribution the male age-specific fecundity function $m^*(x)$, as defined by equation (1.7), must be independent of time. A *male* reproductive function $k^*(x) = l^*(x)m^*(x)$ can thus be defined for the stable age-distribution; application of an argument similar to that leading to equation (1.17) for the female population gives the relation (for $t \geqq d^*$)

$$B_m(t) = \sum_{x=b^*}^{d^*} B_m(t-x)k^*(x)$$

(Modifications similar to those made in deriving equation (1.19) can be used to obtain an expression for $B_m(t)$ when $t < d^*$.) Use of the renewal theory applied to the female population process on pp. 27–9 gives us the characteristic equation

$$\sum_{x=b^*}^{d^*} \lambda^{-x}k^*(x) = 1 \tag{1.63}$$

From the above considerations, the leading root of this equation must be λ_1. Equations (1.32) and (1.63) thus relate the male and female demographic parameters for the stable age-distribution. The values of $m^*(x)$ used for obtaining $k^*(x)$ must, of course, be found either from empirical data on the male age-specific fecundity schedule for the stable age-distribution, or by detailed specification of the rules of mating with respect to male and female ages. Generation times, reproductive values etc. can be defined using $m^*(x)$ and $k^*(x)$, just as for the female population.

Plants with seed dormancy.   The population projection matrix in equation (1.14) can be handled, for the case of time-independent demographic parameters, by matrix methods similar to those used on pp. 24–6 for the female population. The characteristic equation is of a rather more complex form than equation (1.32), however, and it is simpler to derive it using the difference equation approach. Let us define the following quantities; $l'(x)$ $(x = 1, 2, \ldots d'+1)$ is the probability that a seed survives to be aged x, and then germinates; $l(x)$ $(x = 0, 1, \ldots d-1)$ is the probability that a plant survives x years from the time of germination. We have

$$l'(1) = P''(0) \tag{1.64a}$$

$$l'(x) = P''(x-1) \prod_{y=2}^{x} P'(x-y) \quad (x = 2, 3, \ldots d'+1) \tag{1.64b}$$

$$l(0) = 1 \tag{1.64c}$$

$$l(x) = \prod_{y=0}^{x-1} P(x-y) \quad (x = 1, 2, \ldots d-1) \tag{1.64d}$$

The net expected number of progeny seeds ascribable to a seed x years after it was itself formed is thus

$$K(x) = \sum_{y=0}^{x-1} l'(x-y)l(y)M(y+1) \quad (x = 1, 2, \ldots d+d') \qquad (1.65)$$

If we define $B(t)$ as the total number of seeds produced by the population during the flowering season of year t, we have (for $t \geq d+d'$)

$$B(t) = \sum_{x=1}^{d+d'} B(t-x)K(x) \qquad (1.66)$$

Applying the renewal theory results, the asymptotic rate of increase of the population is thus the leading root λ_1 of the equation

$$\sum_{x=1}^{d+d'} \lambda^{-x}K(x) = 1 \qquad (1.67)$$

Provided that $K(x)$ is not periodic, λ_1 is real and positive. As the population approaches its asymptotic growth-rate, the age-structures of both the seed and adult plant populations must stabilise.

1.4   Time-dependent and density-dependent demographic parameters

The solution of the population process studied in the preceding section of this chapter is based on the assumption that the demographic parameters of the population are independent of time. This is clearly at best an approximation to the behaviour of a real population, and it obviously is important to examine the consequences of relaxing this assumption. It is useful to distinguish between two ways in which time-dependence of the demographic parameters can arise; factors of the external environment which affect survival or fecundity may vary over time (*time-dependence* in the strict sense), or changes in the size and age-composition of the population may cause changes in the demographic parameters (*density-dependence*). Of course, both of these processes may operate at the same time in a real population, but it is simplest to treat them as alternatives. We begin by discussing the case of time-dependent demographic parameters, and then go on to consider density-dependence.

1.4.1 *Time-dependent demographic parameters*

This situation may be further sub-divided into two. There is firstly the case of a deterministically varying environment, where we can follow a population which is exposed to some definite pattern of variation over time in its demographic parameters. Secondly, there is the case of a stochastically varying environment, where the successive sets of demographic parameters to which a population is exposed are governed by some probability law. As we shall see, it is not possible in either of these cases to obtain useful analytical solutions for the age-structure and growth-rate of the population at a given time, as was possible with time-independence. It *is* possible to prove, however, that the initial age-structure of the population eventually ceases to influence the age-structure and growth-rate of the population. This is the so-called *weak ergodicity* property of the population process, which may be contrasted with the strong ergodicity property of a population with fixed demographic parameters (p. 31); in that case, the asymptotic state is characterised by a constant, calculable age-distribution and growth-rate. For simplicity, we shall consider only the female population in the following analysis.

Deterministically varying environments. Consider first of all the case of a discrete age-class population. Its state at time t is characterised by the vector of number of individuals in different age-classes, $n(t)$, and by the Leslie matrix for time t, $L(t)$ (cf. pp. 3–6). Using equations (1.5) and (1.6), we can express $n(t)$ in terms of the initial population vector $n(0)$, and the sequence of Leslie matrices between times 0 and t

$$n(t) = n(0) \prod_{u=0}^{t-1} L(u) \tag{1.68}$$

This result is trivial and does not provide in itself any insight into the properties of the population process, although it can be used for numerical calculations of population trajectories. Lopez (1961) used it, however, to prove that the age-structure of the population eventually becomes independent of the initial vector $n(0)$. Let the elements of the vectors $n_1(t)$ and $n_2(t)$ describe two populations with two different initial vectors, $n_1(0)$ and $n_2(0)$, but which are exposed to the same set of Leslie matrices; then, for large t, the following equation is satisfied

$$\frac{n_1(x, t)}{n_2(x, t)} \sim \frac{n_1(y, t)}{n_2(y, t)} \quad (x \neq y) \tag{1.69a}$$

If we use the fact that the demographic parameters are identical for the two populations at any given time, it is easily verified that this relation implies that the two population vectors satisfy the equation

$$\boldsymbol{n}_2(t) \sim C\boldsymbol{n}_1(t) \tag{1.69b}$$

where C is a constant whose value depends on the initial vectors $\boldsymbol{n}_1(0)$ and $\boldsymbol{n}_2(0)$. This is turn implies that the geometric growth-rate of the population at time t, as measured by the relative values of (say) $n(1, t+1)$ and $n(1, t)$, becomes dependent only on t, and is independent of the initial conditions.

The following conditions are sufficient for the validity of this theorem. The non-zero elements of L are bounded, and greater than some fixed number $\varepsilon > 0$. There are two values of x, x_1 and x_2, with greatest common divisor of one, such that $f(x_1, t)$ and $f(x_2, t)$ are non-zero for all values of t. (This is related to the requirement of non-periodicity for the time-dependent case, discussed on p. 26.) The survival probabilities $P(x, t)$ are non-zero, as well as at least some of the elements of the initial population vector $\boldsymbol{n}(0)$. The proof requires fairly lengthy matrix manipulations, which are beyond the scope of this book. The interested reader should consult Lopez (1961), Pollard (1973, pp. 51–5) for details of the proof. Seneta (1973, pp. 69–77) has proved this theorem for the case of non-negative matrices of more general structure than the Leslie matrix; his results are used in section 2.2.

A theorem of the same sort exists for continuous-time populations, and was first stated and proved by Norton (1928), in a paper which has largely been overlooked by workers in demography. Norton assumed that the reproductive function $k(x, t)$ in equation (1.27) is some defined function of t as well as x, due to temporal variation in environmental factors affecting survival or fecundity. He assumed that there were fixed upper and lower limits, b and d, to the values of x for which $k(x, t)$ is non-zero; and that $k(x, t)$ is an integrable function of x and t. (Continuity of $k(x, t)$ with respect to both x and t is sufficient but not necessary for the validity of Norton's results.) Provided that there are at least some individuals of reproductive or pre-reproductive age at time $t = 0$, Norton showed that:

> given two communities in which there is the same birth-rate and death-rate [age-specific fecundities and death-rates in our terminology], the number of births in one tends ... to a constant proportion of the number of births at that moment in the other ... Finally, if the birth-rate and death-rate are given – and provided the conditions stated are fulfilled – the eventual age-distribution is determinable; it will vary, of course, from moment to moment, but is approximately the same at the same time in all communities in which the birth- and death-rates have their given values; in particular, it will not depend on the initial age-distribution (Norton, 1928, p. 20).

As in the case of time-independent demographic parameters, the properties of the discrete age-class and continuous-time models are parallel. The general conclusion is that the age-structure and rate of growth of a population subjected to demographic parameters which vary with time tend asymptotically to values which are independent of the initial state of the population. The speed of this convergence is difficult to study analytically; some numerical studies have been made by Kim and Sykes (1976), who conclude that convergence takes place over a few generations for sets of data drawn from human populations. These results are of some importance in population genetics, as we shall in Chapters 2 and 4.

Stochastically varying environments. Cohen (1976, 1977a, 1977b) has generalised these conclusions to the case of a population exposed to stochastic variation in the demographic parameters. He considers a discrete age-class model described by Leslie matrices of the same structure assumed by Lopez (1961); the sequence of Leslie matrices applied to a population is assumed to be sampled from an ergodic Markov chain (i.e. the probability law relating the state of L at two successive times is such that the distribution of $L(t)$ becomes independent of t for sufficiently large t). Under these conditions, the age-structure and growth-rate of the population exhibit *stochastic ergodicity*; the probability distribution of the age-structure of the population eventually becomes independent of the initial vector

$n(0)$, and of the initial distribution of the demographic parameters. The details of the assumptions and methods involved in the proof of this result can be found in the reference cited above. Cohen (1977*b*) gives an example of a simulation of this type of population process, which illustrates the convergence of the probability distribution of the age-structure to its limiting state.

1.4.2   *Density-dependent demographic parameters*

It is only exceptionally that organisms find themselves in an environment where there is no check to their continued increase in numbers due to the operation of factors which increase mortality or reduce fecundity, in response to increasing population density in a local area (density-dependent factors). Although the nature of such checks to population growth varies from species to species, and the full details of the mechanism of regulation of population density have probably never been worked out for even one species, there can be little doubt of their general importance in ecology (Lack, 1954; Nicholson, 1957, 1960). One of the standard texts on ecology, such as Krebs (1972) or Ricklefs (1973), may be consulted for more information about density-dependent factors in natural populations. Our concern here is to examine how density-dependent demographic parameters can be incorporated into models of the sort outlined earlier in this chapter (pp. 3–15). It is useful to start by considering the properties of models of density-dependent populations which neglect age-structure.

Density-dependence in populations without age-structure. The classical approach to modelling a density-dependent population is to write down a differential equation which describes the rate of increase in the number of individuals, n, in a local population

$$\mathrm{d}n/\mathrm{d}t = ng(n) \tag{1.70}$$

where $g(n)$ is a continuous, decreasing function of n. As we have done earlier, it is best to regard n as the number of females only. The simplest and most widely used form for $g(n)$ is that of the *logistic growth* or *Pearl–Verhulst* equation

$$g(n) = \alpha - \beta n \tag{1.71}$$

The expression which results if equation (1.71) is substituted into (1.70) is readily integrated to give n as a function of time

$$n(t) = \frac{\alpha n(0)}{\beta n(0) + [\alpha - \beta n(0)] e^{-\alpha t}}$$ (1.72)

If the population is started with a population size close to zero, it increases initially at a logarithmic rate of approximately α; provided n is initially positive, the population converges smoothly towards a stationary population size of

$$\hat{n} = \alpha/\beta$$ (1.73)

$\hat{n}$ is usually referred to as the *carrying-capacity* of the population. The logistic equation provides a reasonably good empirical fit to the form of population growth of model laboratory populations, such as population cages of *Drosophila* (e.g. Buzzati-Traverso, 1955).

More complicated forms for $g(n)$ can be studied in a similar way (May, 1973). If $g(n)$ is a *strictly* decreasing function of n, it is obvious from equation (1.70) that there is only one carrying-capacity or stationary population size $\hat{n} > 0$, which must satisfy $g(n) = 0$. It is not always possible to integrate equation (1.70) directly to obtain n as an explicit function of t. The convergence to the carrying-capacity can be studied in such a case by examining the *local stability* of the equilibrium, $\hat{n}$. Applying Taylor's theorem to equation (1.70), for n close to $\hat{n}$, and noting that $g(\hat{n}) = 0$, we obtain

$$\frac{dn}{dt} \cong (n - \hat{n}) \left(\frac{\partial g}{\partial n} \right)_{\hat{n}}$$

i.e.

$$n(t) - \hat{n} \cong [n(0) - \hat{n}] \exp \left[t \left(\frac{\partial g}{\partial n} \right)_{\hat{n}} \right]$$ (1.74)

Since g is a decreasing function of n, $(\partial g/\partial n)_{\hat{n}}$ is negative, so that the exponential term decays to zero with increasing time. The equilibrium is therefore locally stable; a small perturbation away from the carrying-capacity results in a return towards equilibrium.

Equation (1.70) is clearly a highly over-simplified representation of the dynamics of a population. It is strictly valid only if the individuals have death-rates $\mu(x, t)$ and fecundities $m(x, t)$ which are independent of age (x), but depend on total population size. In order to avoid the loss of realism inherent in applying such equations to age-structured populations reproducing in continuous time, theoretical ecologists have developed similar models of density dependence for populations with completely discrete generations. These provide reasonably accurate descriptions of, for example, annual plants and many semelparous insect species of economic importance, which live only one year. Equation (1.70) can be replaced by the related first-order difference equation

$$n(t) = n(t-1)w[n(t-1)] \qquad (1.75)$$

where $n(t)$ is the number of adult females alive in generation t, and w is the expectation of female offspring (counted at maturity) of an adult female in the preceding generation. A plausible and simple form for $w(n)$, analogous to the logistic form for the continuous-time case, is

$$w(n) = \alpha \, e^{-\beta n} \qquad (1.76)$$

The carrying-capacity $\hat{n}$ now satisfies the equation

$$w(n) = 1 \qquad (1.77)$$

If $g(n)$ is a strictly decreasing function of n, there is a unique solution to this equation. The dynamics of the population in the neighbourhood of the equilibrium can again be studied by a local analysis. We have

$$n(t) - \hat{n} \cong [n(t-1) - \hat{n}]\left[1 + \hat{n}\left(\frac{\partial w}{\partial n}\right)_{\hat{n}} \right] \qquad (1.78)$$

If $-2 < \hat{n}(\partial w/\partial n)_{\hat{n}} < 0$, the equilibrium is locally stable. If $\hat{n}(\partial w/\partial n)_{\hat{n}} < -2$, the population responds to small perturbations away from equilibrium by oscillations of progressively increasing magnitude across $\hat{n}$, i.e. it is locally unstable. If $w(n)$ is of the form given by equation (1.76), for example, we have $\hat{n} = (\ln \alpha)/\beta$ and $\hat{n}(\partial w/\partial n)_{\hat{n}} = -\ln \alpha$, so that the equilibrium is locally stable for $\ln \alpha < 2$, and unstable if $\ln \alpha > 2$. Too great a sensitivity of the population growth-rate to population density thus leads to instability. But this stability analysis is valid only for the neighbourhood of an equilibrium; it is obvious that a population close to $n = 0$ must tend to increase in number, and a population with very large n will tend to decrease, provided that $w(n)$ is a decreasing function of n. This suggests that the local instability of the equilibrium actually corresponds to some kind of bounded fluctuations in population size. This was apparently first pointed out in an ecological context by Moran (1950) and, independently, by Ricker (1954).

More detailed analyses of the dynamical behaviour of expressions of the form of equation (1.75) have recently considerably deepened our understanding of the complex dynamics which are possible when there is formally a locally unstable equilibrium (reviewed by May and Oster, 1976). Consider, for example, the behaviour of the population model expressed by equation (1.76) with different parameter values. If $\hat{n}$ is kept fixed, so that $\ln \alpha$ and β are varied proportionately, it is found that the population converges to $\hat{n}$ for $\ln \alpha < 2$. When $\ln \alpha$ is slightly greater than 2, the population approaches a *limit cycle* of period 2, i.e. the population trajectory settles into a state where n switches each generation between two alternative values, one above $\hat{n}$ and the other below. If $\ln \alpha$ and β are further increased, this limit cycle is replaced by one with period 4 (the population approaches a state in which it switches between four successive values of n), which is in turn replaced by one of period 8, etc. The dynamical behaviour of the population thus experiences a succession of *bifurcations* such that, with increasing sensitivity to population density, successive stable limit cycles of period $4, 8, \ldots 2^m, \ldots$ are generated. It can be shown that the pattern of bifurcation behaviour depends only on α, so that two populations with different $\hat{n}$ but the same α will have the same type of limit cycle (May and Oster, 1976). (The actual values of n through which the population passes are, of course, dependent on β as well as α.) When $\ln \alpha$ exceeds 2.6924, there is an infinite number of limit cycles of differing periods; there is also an uncountable set of initial conditions for which the population trajectories are aperiodic, so that the pattern of population size with respect to time never repeats itself. This is the so-called region of *chaos*. A population in this region will behave in an essentially unpredictable fashion, with periods of low population size succeeded by sudden increases in number, followed by rapid return to a low level once more.

As is obvious from the simple method described above, cyclic or chaotic behaviour depends on the species having a high reproductive capacity, so that α is sufficiently large. If the rate of population growth at low density is small, then the population will settle down to a stable equilibrium level, $\hat{n}$, as given by equation (1.77). Hassell, Lawton & May (1976) have analysed data on several species, and concluded that most seem to be in the region of stability or 2-point or 4-point limit cycles, so that chaotic phenomena are perhaps infrequent in natural populations.

Models of density-dependence with age-structure. The complex behaviour of discrete-generation populations, revealed by the studies summarised above, suggests that it is important to examine carefully the properties of age-structured populations with density-dependence. Various special cases have been studied numerically by Leslie (1948), Pennycuick, Compton & Beckingham (1968), Usher (1972), Beddington (1974), Smouse and Weiss (1975), and others. Analytical treatments have been given by Gurtin and Mac-Camy (1974), Oster and Takahashi (1974), Oster (1976), Rorres (1976) and Guckenheimer, Oster and Ipaktchi (1977). We will first consider some general properties of density-dependent regulation in age-structured populations, and then go on to discuss the dynamical properties of some special cases.

It is clear, both on theoretical grounds and from the results of empirical studies, that density-dependent regulation in an age-structured population does not usually involve a response of the demographic parameters to the total number of individuals present at one time. More usually, the density-dependent components of the demographic parameters respond only to the numbers of individuals in a restricted sub-group of the population. Such a sub-group has been called the *critical age-group* by Charlesworth (1972). For example, in many bird species density-dependence seems to occur through limited availability of food for nestlings (Lack, 1954, 1966), so that the survival of nestlings decreases with an increase in the number of nestlings in a given area. Similarly, seedling survival in plants is affected by competition between seedlings for light and space, resulting in negative dependence of seedling survival on the numbers of seedlings (Harper and White,

1974). The critical age-group in these cases is composed simply of the nestlings or seedlings, respectively. Nicholson (1957, 1960) describes experimental populations of the sheep blowfly *Lucilia cuprina* in which fecundity of adult females was controlled by competition for a limited food supply. In this case, the critical age-group consists of the entire adult female population.

The simplest type of reasonably realistic model of density-dependence in a discrete age-class model represents the age-specific survival probabilities or fecundities, for one or more age-classes, as decreasing functions of the number of individuals in the critical age-group. Let this number be $N(t)$ at time t. Consider, for example, the case of a species which starts reproduction b years after birth, and in which juvenile survival depends on the total number of juvenile individuals. For simplicity, only the female population is considered. We can write the survival probability for a juvenile aged x at time t as

$$P(x, t) = P^{(I)}(x)P^{(D)}[x, N(t)] \quad (x = 0, 1, \dots b-1)$$

where

$$N(t) = \sum_{x=1}^{b-1} n(x, t)$$

$P^{(I)}(x)$ is the density-independent component of juvenile survival for age-class x; $P^{(D)}[x, N(t)]$ is the density-dependent component, which is assumed to be a decreasing function of $N(t)$.

The population process in this case can be represented by a Leslie matrix with density-dependent survival probabilities $P(x, t)$ for ages $x = 1, 2, \dots b-1$, and density-dependent net fecundities $f(x, t) = m(x)P(0, t)$. Alternatively, we can represent the process by a difference equation of the form of equations (1.17) or (1.19). We can write

$$k(x, t) = k^{(I)}(x)k^{(D)}(x, t) \tag{1.79a}$$

where

$$k^{(I)}(x) = m(x) \prod_{y=0}^{b-1} P^{(I)}(y) \prod_{y'=b}^{x-1} P(y') \tag{1.79b}$$

$$k^{(D)}(x, t) = m(x) \prod_{y=0}^{b-1} P^{(D)}[y, N(t+y-x)] \tag{1.79c}$$

The dependence of the reproductive function on the values of $N(t)$ over a set of past times, as expressed in equation $(1.79c)$, is typical of what happens with density-dependence in an age-structured population, and greatly complicates the analysis of its dynamics. In general, we can represent it by writing

$$k(x, t) = k(x, N_T) \tag{1.80}$$

where N_T represents the set of values of $N(t)$ over some set T of past times. As in the above example, the actual set of times included in T may depend on x. The population process is then described in general terms (for times $t \geq d$) by the equations

$$N(t) = \sum_{x \in S} B_f(t-x)l(x, N_T) \tag{1.81a}$$

$$B_f(t) = \sum_{x=b}^{d} B_f(t-x)k(x, N_T) \tag{1.81b}$$

S in equation $(1.81a)$ represents the set of age-classes which comprise the critical age-group.

Appropriate modifications can be made for $t < d$ (equation (1.19)). An equivalent matrix formulation can obviously be written down, but the difference equation method of description is convenient for dealing with the problem of selection in density-dependent populations, as will be discussed in Chapters 3 and 4. The extension to the corresponding continuous-time models is straightforward and will not be given here.

Stationary populations. By analogy with the results for discrete-generation populations, we would expect that an age-structured population with a sufficiently low reproductive capacity at low population density should converge towards an equilibrium state with a stationary number of individuals in each age-class. In such a stationary population, equation $(1.81b)$ gives us the relation

$$\sum_{x} k(x, N_T) = 1 \tag{1.82a}$$

By definition, the number of individuals in the critical age-group is unchanging, so that each number in the set N_T is constant at the equilibrium value of $N(t)$, $\hat{N}$ say. This corresponds to the carrying-capacity of the population in the simple models described above (pp. 51–5). The value of $\hat{N}$ can be determined from this equation by the following procedure. We can imagine a set of populations held at different stationary states by some artificial means. Let the number of individuals in the critical age-group in one such imaginary population be N, say. If the density-dependent components of the demographic parameters are *strictly* decreasing functions of the number of individuals in the critical age-group, it follows that, in such a stationary population, the net reproduction rate is a strictly decreasing function of N. $\hat{N}$ is thus given as the *unique* root of the equation

$$\sum_x k(x, N) = 1 \tag{1.82b}$$

This equation is analogous to equation (1.45), which gives the intrinsic rate of increase in the density-independent case. As in that case, the solution usually has to be obtained by numerical iteration of the equation.

The uniqueness of $\hat{N}$, which follows from the particular model of density-dependence just described, is useful in analysing the mathematical properties of the system, especially in connection with the theory of selection with density-dependence, to be discussed in Chapters 3 and 4. It should be borne in mind that, although this class of model probably provides a fairly realistic description of the process of density regulation in many situations, other types of model with more complex properties may apply to individual cases. For example, different age-classes within the critical age-group may contribute differentially to the impact of density on the demographic parameters, or the decreasing properties of the latter as functions of $N(t)$ may hold only for a certain range of population size. Beddington (1974) discusses an example of the second sort, which results in two alternative equilibria for the same set of demographic parameters.

Non-equilibrium behaviour. A detailed analysis of the non-equilibrium behaviour of these models usually presents considerable mathematical difficulties, and computer studies have frequently been resorted to. Oster (1976) reviews the general mathematical aspects of the population dynamics resulting from density-dependence with age-structure, as well as some studies of individual cases. In order to illustrate the main consequences of density-dependence we shall concentrate here on one particular model. We assume that the critical age-group is the set of all adult females, and that the fecundity of adult females at time t is a strictly decreasing function of $N(t)$, the number of individuals in the critical age-group. For simplicity, we assume that the fecundity of each adult age-class is affected to the same extent by density, so that we can write

$$m(x, t) = m^{(I)}(x)m^{(D)}[N(t)] \qquad (1.83a)$$

where

$$N(t) = \sum_{x=b}^{d} B_f(t-x)l(x) \qquad (1.83b)$$

$m^{(D)}$ is a strictly decreasing function of N, and takes a maximum value of unity when $N = 0$. The population trajectory can be obtained by substituting from these equations into equations (1.81), noting that in this case $k(x, N_T) = l(x)m(x, t)$. The carrying-capacity $\hat{N}$ is given by substituting into equation (1.82b). It is convenient to use the following simple form (analogous to equation (1.76) for the discrete-generation case) to represent the dependence of $m^{(D)}$ on density

$$m^{(D)}[N(t)] = e^{-\beta N} \qquad (1.84)$$

$\hat{N}$ is then given by the equation

$$\hat{N} = (\ln R^{(I)})/\beta \qquad (1.85)$$

where $R^{(I)} = \sum l(x)m^{(I)}(x)$ is the net reproduction rate of the population at low population density, and obviously plays a similar role to α in the discrete-generation case. (The rate of growth achieved asymptotically by the population at low density may be obtained by substituting $l(x)m^{(I)}(x)$ for $k(x)$ in equation (1.45).)

Some insight into the dynamics of this system can be obtained by carrying out a local stability analysis around the equilibrium point, on lines similar to that yielding equation (1.78). Let $\delta B_f(t)$ and $\delta m^{(D)}(t)$ be the deviations at time t of B_f and $m^{(D)}$ from their respective equilibrium values, $\hat{B}_f = \hat{N}/\sum_{x=b}^{d} l(x)$ and $m^{(D)}$ $(\hat{N})$. Neglecting second-order terms in these quantities, equations (1.81) give (for $t \geqq d$)

$$\delta B_f(t) = \sum_{x=b}^{d} \delta B_f(t-x) k(x, \hat{N}) + \hat{B}_f \delta m^{(D)}(t) \sum_{x=b}^{d} l(x) m^{(I)}(x) \quad (1.86)$$

But

$$\delta m^{(D)}(t) = \left(\frac{\partial m^{(D)}}{\partial N}\right)_{\hat{N}} \sum_{x=b}^{d} \delta B_f(t-x) l(x)$$

$$= -\beta m^{(D)}(\hat{N}) \sum_{x=b}^{d} \delta B_f(t-x) l(x)$$

Substituting from this into equation (1.86), and noting that $m^{(D)}(\hat{N}) \sum l(x) m^{(I)}(x) = \sum k(x, \hat{N}) = 1$, we obtain

$$\delta B_f(t) = \sum_{x=b}^{d} \delta B_f(t-x)[k(x, \hat{N}) - \beta \hat{B}_f l(x)] \quad (1.87)$$

This is a renewal equation similar to that studied earlier (pp. 27–9). It follows that the equilibrium is locally stable (δB_f tends to zero) when the leading root of the following equation has modulus less than unity

$$\sum_{x=b}^{d} \lambda^{-x}[k(x, \hat{N}) - \beta \hat{B}_f l(x)] = 1 \quad (1.88a)$$

Further analysis is simplified if we assume that $m^{(I)}(x)$ as well as $m^{(D)}$ is independent of x for $b \leq x \leq d$, so that adult fecundity is independent of age. Exploiting the equilibrium relations given by equation (1.85), equation (1.88a) can then be simplified to the following form

$$\left[\sum_{x=b}^{d} \lambda^{-x} l(x)\right][1 - \ln R^{(I)}] = \sum_{x=b}^{d} l(x) \quad (1.88b)$$

If $\ln R^{(I)} < 1$, the argument used on pp. 25–7 establishes that the leading root of equation (1.88b) is real, positive and less than unity. In such cases, the equilibrium is locally stable, so that small perturbations are followed by a return towards equilibrium. Computer calculations of population trajectories show that, in all cases studied, there is convergence to the equilibrium from any starting point whenever this condition is satisfied.

A *necessary* condition for local instability of the equilibrium is that the reproductive capacity of the species is such that the natural logarithm of the net reproduction rate at low population density exceeds unity. A *sufficient* condition is less easy to find, unless we make further assumptions about the life-history. For example, let us assume that reproduction starts in age-class 1, and that there is a constant survival probability P for each age-class. (This approximates the life-history of many species of small birds and mammals, as we saw in section 1.2.3.) We then have

$$\sum_{x=b}^{d} \lambda^{-x} l(x) = \sum_{x=1}^{d} \lambda^{-x} P^x = \lambda^{-1} P(1 - \lambda^{-d} P^d)/(1 - \lambda^{-1} P)$$

so that

$$\sum_{x=1}^{d} l(x) = P(1 - P^d)/(1 - P)$$

For large d, the terms $\lambda^{-d} P^d$ and P^d can be neglected (assuming that $\lambda > P$), so that equation (1.88b) becomes

$$\lambda = P + (1 - P)(1 - \ln R^{(I)}) \tag{1.89}$$

Equation (1.89) gives the following necessary and sufficient condition for local instability of the equilibrium ($\lambda < -1$)

$$\ln R^{(I)} > 2/(1 - P) \tag{1.90}$$

By analogy with the results for the discrete-generation case (pp. 53–5 above), we would expect a population which satisfies this inequality to display periodic or chaotic behaviour instead of settling down to a stationary state. Computer calculations of population trajectories confirm this. A population for which the inequality is only just satisfied approaches a stable limit cycle of period 2 about $\hat{N}$; populations with higher reproduction rates exhibit period 4 limit cycles, etc.; chaotic phenomena are found with sufficiently high $R^{(I)}$ values. It is obvious from equation (1.90) that $R^{(I)}$ must exceed

$e^2 \cong 7.39$ for instability; the higher the survival probability P, the higher $R^{(I)}$ must be (e.g. if $P = 0.5$, it is necessary that $R^{(I)} > 54.6$). Even if the instability condition is satisfied, computer calculations show that the amplitude of the population fluctuations is usually small. The instability phenomena are, therefore, less marked than with the corresponding discrete-generation model, and it seems likely that populations of birds and mammals with this kind of life-history will tend to settle down to a stationary state.

A somewhat different picture emerges if we consider a life-history in which there is a long phase of pre-reproductive life, as in many insect species. This is obviously closer to the discrete-generation case, and one might therefore expect it to be easier to obtain instability than in the case just discussed. This conclusion is confirmed by computer calculations of population trajectories. A classification of the instability phenomena associated with increasing $R^{(I)}$ is shown in Table 1.5, for a case in which reproduction

Table 1.5. *Stability behaviour with density-dependence*

$m^{(I)}$	β ($\times 10^{-3}$)	$R^{(I)}$	Stability properties
68	1.38	3.98	Convergence to $\hat{N}$
136	2.08	7.97	Convergence to $\hat{N}$
153	2.19	8.94	Period 10 limit cycle
170	2.30	9.96	Chaos (low-amplitude fluctuations)
850	3.91	49.95	Chaos (high-amplitude fluctuations)
1700	4.60	99.61	Chaos (high-amplitude fluctuations)

This is a model with eight age-classes, but where reproduction starts in age-class 5. The fecundity of each reproductive age-class is the same ($m^{(I)}$ and $m^{(D)}$ are both independent of age). The survival probability for each age-class is 0.5. The parameters are chosen such that the equilibrium value of N (number of adult females) is 10^3.

starts in age-class 5 and where the survival probability and fecundity of each age-class are independent of age. It is evident that instability occurs at a considerably lower fecundity level than in the previous example, although a fairly high value of $R^{(I)}$ is still necessary. The chaotic regime in this case may result in nearly periodic population fluctuations of a length ($\cong$ ten-intervals) somewhat greater than the total number of age-classes. An example is shown in Figure 1.3. The periodicity of ten time-intervals is evidently related to the fact that reproduction is initiated five age-classes after conception, so that a delay of five time-intervals results between any reduction in the number of eggs laid by the adult population, in response to an excessive number of adults, and the consequent reduction in the number of adults. The periodicity occasionally breaks down, and is replaced by apparently random fluctutations which are then followed by re-establishment of the periodic changes. This parallels quite closely the results of the classic experiments of Nicholson (1957, 1960) on *Lucilia cuprina*, where violent, nearly periodic fluctuations in population size were produced in artificial populations subjected to density limitation by competition of adults or

Figure 1.3. The fluctuations in the numbers of individuals in a population subject to density-dependent regulation of the type shown in Table 1.5, bottom line. The numbers on the ordinates represent thousands of individuals. The initial population consisted of adults only, with 1000 in each age-class.

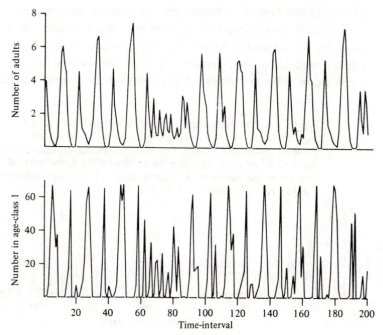

larvae for food (Figure 1.4). The model described here corresponds roughly to the experiments in which the fecundity of female flies was limited by a restricted food supply (because an adequate amount of food is necessary for the formation of eggs by the adult female flies). The periodicity observed by Nicholson was of the order of 30 days, and his data show that the time from egg to reproductive maturity is around 15 days, in agreement with the relationships found in the computer model. Oster (1976) has constructed and analysed a more sophisticated model, which closely predicts the pattern of population behaviour observed in Nicholson's experiments. It is remarkable that the appearance of random fluctuations superimposed on a basically periodic process can be produced in a purely deterministic model.

The results discussed here demonstrate that age-structured populations subject to density-dependent limitation of population growth may show cyclical or chaotic fluctuations in numbers, provided that the reproductive capacity of the species is sufficiently high. Organisms with relatively low fecundity may be expected to approach a stationary population size under density-dependence. The instability behaviour found with the discrete age-class model may occur in continuous-time populations, under suitable conditions; the local stability analysis leading to equation (1.88) could equally well have been carried out with the integral equation formulation. It is a matter for future research to determine whether

Figure 1.4. The fluctuations in numbers of individuals in a cage population of *L. cuprina*, regulated by a limited but constant adult food supply. The continuous line represents the numbers of adults in the cage on successive days; the vertical lines show the numbers of adults emerging from eggs laid over 2 days. (After Nicholson, 1957, Fig. 8I.)

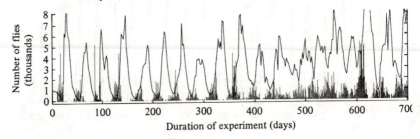

or not these instability phenomena are important agents in contributing to cyclic population phenomena in nature (cf. Oster and Takahashi, 1974; Hassell *et al.*, 1976).

1.5   Systems of several populations

We have so far considered only the case of a single, genetically homogeneous population. In practice, of course, a species consists of a mixture of different genotypes, and exists as a member of a community of species which interact with each other by competition, predation, parasitism, mutualism, etc. Most ecological models of interacting species have ignored the problem of age-structure, although models of predator–prey interactions, for example, have been studied by Maynard Smith and Slatkin (1973), Auslander, Oster and Huffaker (1974) and Beddington and Free (1976). It is clear from these studies that incorporating age-structure may introduce novel features into the behaviour of multi-species systems. As this book is concerned mainly with the consequences of genetic differences within a species, the problem of ecological interactions between species will be considered only briefly. An asexually reproducing population which contains a number of genetically distinct clones is formally equivalent to a system of different, non-interbreeding species. We will examine here the simplest kind of system of this sort: one in which the different sub-populations (asexual genotypes or species) have different sets of demographic parameters which remain constant over time, and are unaffected by the composition of the sytem. This obviously excludes the standard types of ecological interaction mentioned above, and is therefore most relevant to the process of selection within an asexually reproducing population. The models explored below were originally investigated by Pollak and Kempthorne (1970, 1971).

1.5.1   *Continuous-time populations*

It is convenient to examine the continuous-time model first. An asexual population with m distinct genotypes $A_1, A_2, \ldots A_m$ can be regarded as composed of m separate sub-populations, each with its own set of demographic parameters. The sub-population A_i is characterised by a reproductive function $k_i(x)$, with which an intrinsic rate r_i is associated (r_i is obtained by applying equation (1.42) to $k_i(x)$). In the case of an asexual population, there is obviously no need to distinguish between the sexes in calculating fecundities; if the system is regarded as representing a set of sexual but non-interbreeding species, we can follow our earlier practice, and treat the female part of each population only.

Let $n_i(x, t)dx$ be the number of A_i individuals aged between x and $x + dx$ at time t; the total number of A_i individuals is $n_i(t) = \int_0^d n_i(x, t) \, dx$, and the total number of individuals in the system is $n(t) = \sum_i n_i(t)$. The theory developed on pp. 30–1 tells us that, for sufficiently large t, we can write

$$\frac{dn_i}{dt} = r_i n_i \tag{1.91}$$

The state of the system is most conveniently represented in terms of the frequencies of the different types, and the total number of individuals n. The frequency of A_i at time t is defined as $p_i(t) = n_i(t)/n(t)$; it is also useful to define the mean intrinsic rate of increase, taken over all types, as

$$\bar{r}(t) = \sum_i p_i(t) r_i \tag{1.92}$$

Using equations (1.91) and (1.92), we find (with some rearranging) that

$$\frac{dn}{dt} = \bar{r}n \tag{1.93a}$$

and

$$\frac{dp_i}{dt} = p_i(r_i - \bar{r}) \tag{1.93b}$$

Substituting from equations (1.93) into (1.92), we obtain an expression for the rate of change of $\bar{r}$:

$$\frac{d\bar{r}}{dt} = \sum_i p_i(r_i - \bar{r})^2 \tag{1.94}$$

This states that the rate of change of the population growth-rate is, for large t, equal to the variance among types in the intrinsic rate of increase. This is the version of Fisher's Fundamental Theorem of natural selection appropriate for an asexual population (cf. Fisher, 1930); the derivation of a version of this theorem for sexually reproducing populations will be discussed in section 4.2. Equation (1.93*b*) tells us that, not surprisingly, the type with the highest intrinsic rate of increase tends to displace the others from the population; equation (1.94) implies that this process is accompanied by a simultaneous improvement in the growth-rate of the population as a whole.

The use of reproductive value. The above results are obviously valid only for *t* taken sufficiently large that each sub-population has approached its stable age-distribution and rate of growth. Results which are valid for arbitrary *t* can be obtained by applying the concept of reproductive value, which was introduced earlier in the chapter (pp. 37–41). We saw that the rate of increase of the total reproductive value of a population is equal to its intrinsic rate of increase, regardless of the age-distribution. This result can be applied to each sub-population in the present case. We can define the reproductive value, $v_i(x)$, for A_i individuals aged x by using equation (1.59) with the demographic parameters and intrinsic rate of increase of A_i. The total reproductive value of the *i*th sub-population at time *t* is defined as

$$V_i(t) = \int_0^d n_i(x, t) v_i(x) \, \mathrm{d}x$$

The total reproductive value of the population is $V = \sum V_i$. Using equation (1.61), it follows that

$$\frac{\mathrm{d}V_i}{\mathrm{d}t} = r_i V_i \tag{1.95}$$

Weighting each individual by its reproductive value, we obtain the frequency $\tilde{p}_i$ for A_i, using the relation

$$\tilde{p}_i(t) = \int_0^d n_i(x, t) v_i(x) \, \mathrm{d}x / V \tag{1.96}$$

We can also define $\tilde{r}$ as the corresponding mean intrinsic rate of increase, $\tilde{r} = \sum \tilde{p}_i r_i$.

Using the same arguments that were applied to equations (1.91) and (1.92), we obtain

$$\frac{dV}{dt} = \tilde{r}V \tag{1.97a}$$

$$\frac{d\tilde{p}_i}{dt} = \tilde{p}_i(\tilde{r}_i - \tilde{r}) \tag{1.97b}$$

Similarly,

$$\frac{d\tilde{r}}{dt} = \sum_i \tilde{p}_i(\tilde{r}_i - \tilde{r})^2 \tag{1.98}$$

These equations are of exactly the same form as those derived above, but are valid for all t. They resemble the equations which can be derived for continuous-time populations with no age-structure, i.e. in which fecundities and death-rates are independent of age (c.f. Crow and Kimura, 1970, Chapter 1).

1.5.2 *Discrete age-class populations*

Somewhat similar results can be derived for the discrete age-class model. If $n_i(t)$ is the total number of A_i individuals at time t ($n_i(t) = \sum_x n_i(x, t)$), the results derived in section 1.3.1 imply that, for large t, we obtain the following expression for the change in numbers between t and $t-1$.

$$\Delta n_i(t) = n_i(t) - n_i(t-1) = \lambda_i n_i(t-1)$$

where λ_i is the value of λ_1 for A_i (cf. p. 36). Similarly, we can define the total number of individuals as $n = \sum_i n_i$, the frequency of A_i as $p_i = n_i/n$, and the mean rate of increase as $\bar{\lambda} = \sum_i p_i\lambda_i$, just as for the continuous model. Using the same type of argument as for the continuous-time model, we find that, for large t, we have

$$\Delta n = (\bar{\lambda} - 1)n \tag{1.99a}$$

and

$$\Delta p_i = [p_i(\lambda_i - \bar{\lambda})]/\bar{\lambda} \tag{1.99b}$$

Similarly

$$\bar{\lambda}\Delta\bar{\lambda} = \sum_i p_i(\lambda_i - \bar{\lambda})^2 \tag{1.100}$$

These equations are analogous to equations (1.93) and (1.94). If we weight individuals by their reproductive value, we can obtain expressions analogous to equations (1.97) and (1.98). These results resemble those obtainable for an asexual population with discrete generations (cf. Nagylaki, 1977, Chapter 1).

The problem of treating the process of selection in a sexually reproducing population, where individuals can produce offspring whose genotype differs from their own, is considerably more complex than for the asexual case, and forms the subject of Chapters 3 and 4. Before considering this problem, it is desirable to investigate what happens to the genotypic composition of age-structured populations in which there is no selection, i.e. where there is a mixture of genotypes whose demographic parameters are identical. This topic is considered in the next chapter.

2

The genetics of populations without selection

2.1 Introduction

In Chapter 1, the basic concepts of mathematical demography were introduced. We shall start to apply these concepts in the present chapter to the genetics of age-structured populations. The cornerstone of population genetics theory is the Hardy–Weinberg law, which deals with a population which is *not* exposed to the action of any of the standard evolutionary forces: mutation, migration, selection, non-random mating, or random sampling of genes due to finite population size. In such a population, a single autosomal locus reaches an equilibrium with constant gene frequencies, and in which genotypic frequencies are predicted from the gene frequencies by the familiar Hardy–Weinberg formula (Crow and Kimura, 1970, Chapter 2). This result is conventionally derived for the case of a discrete-generation population, where gene frequencies can be shown to remain constant for all time, under the stated assumptions, and in which Hardy–Weinberg frequencies are reached after one generation. There have been a few attempts to extend this result to more general types of populations. Moran (1962, Chapter 2) gave a treatment of a continuous-time model which assumed constancy of gene frequencies for all time. Charlesworth (1974*b*) showed that gene frequencies become asymptotically constant, using a continuous-time model; he gave a proof for a discrete age-class model with time-independent demographic parameters in Jacquard (1974, Chapter 7). This case has also been studied by Gregorius (1976).

In the first part of this chapter, we shall be concerned with the analysis of the approach to genetic equilibrium in the absence of evolutionary factors, using both discrete age-class and continuous-time models. The cases of a single autosomal locus, a sex-linked locus and two autosomal loci will each be studied. The general conclusion reached is that constancy of gene frequencies and Hardy–Weinberg genotypic frequencies are approached only asymptotically, unless the gene frequencies of each age and sex group are equal in the initial population. We might, therefore, expect recently-founded populations to experience transient changes in gene frequencies, which have nothing to do with the conventional forces of evolutionary change.

The other topic of this chapter is the analysis of the consequences of finite population size for evolutionary change in age-structured populations. We shall be particularly concerned with the measurement of effective population number, the parameter which summarises the impact of finite population size on gene frequency change (Wright, 1931, 1938). Numerous publications have been devoted to the definition of effective population number for age-structured populations. Somewhat heuristic treatments have been given by Kimura and Crow (1963), Nei and Imaizumi (1966), Nei (1970), Giesel (1971), and Crow and Kimura (1971). More exact approaches to the problem were introduced by Felsenstein (1971) and Hill (1972), which have been extended by Johnson (1977a, b), Choy and Weir (1978), Hill (1979), and Emigh and Pollak (1979). The account in this chapter is based on the treatments of Felsenstein and Johnson. In addition to effective population number, the probability of fixation of a neutral gene in a finite population will also be discussed. The chapter concludes with a discussion of the frequencies of consanguineous matings in an age-structured population, based on the work of Hajnal (1963).

2.2   Approach to genetic equilibrium

In this section we shall study the process of approach to genetic equilibrium in a number of genetic and demographic models. It is assumed that the population size is sufficiently large that the effect of random sampling of gamete frequencies can be neglected, and that other evolutionary factors such as selection, mutation and migration are also absent. All the genotypes in the population are assumed to have the same demographic parameters. It is simplest to think of these models as representing populations in which mating is at random with respect to genotype, although they can also be used to describe gene frequency change in inbreeding populations, or in populations in which there is assortative mating of a type which does not lead to genotypic differences in fecundity. We shall consider in turn the cases of a single autosomal locus, a sex-linked locus and two autosomal loci.

2.2.1 *An autosomal locus: the discrete age-class case*

Construction of the model. We shall first analyse the case of a population with discrete age-classes, of the type whose demographic properties were examined in some detail in Chapter 1. Suppose that the population under consideration is segregating at an autosomal locus with m alleles $A_1, A_2, \ldots A_m$. There are m^2 possible genotypes if we distinguish between heterozygotes which carry the same pair of alleles, according to the maternal or paternal origins of the alleles in question. We shall adopt the convention that the *ordered genotype* A_iA_j has allele A_i of maternal origin, and A_j of paternal origin.

Consider first the female part of the population. If there are d age-classes, a complete description of its state at time t is provided by the set of m^2 vectors of numbers of females with all possible ordered genotypes, such that the row vector $\boldsymbol{n}_{ij}(t) = [n_{ij}(1, t), n_{ij}(2, t) \ldots n_{ij}(d, t)]$ represents the numbers of A_iA_j females in age-classes $1, 2, \ldots d$. Alternatively, we can write $p_{ij}(x, t)$ for the frequency of A_iA_j among females of age x at time t, and $\boldsymbol{p}_{ij}(t) = [p_{ij}(1, t), p_{ij}(2, t) \ldots p_{ij}(d, t)]$ for the vector of frequencies of A_iA_j across age-classes of females. The vector of numbers of females of all genotypes, $\boldsymbol{n}(t) = \sum_{ij} \boldsymbol{n}_{ij}(t)$, together with $m^2 - 1$ of the frequency vectors can also be used to describe the population. Similar vectors, of dimension d^*, can be defined for the male part of the population. As in Chapter 1, the vectors for males will be distinguished from their female counterparts by asterisks.

An abbreviated description of the system is provided by considering numbers or frequencies of *alleles* rather than genotypes. Thus, we can define the vector $\boldsymbol{n}_i(t) = [n_i(1, t), n_i(2, t), \ldots n_i(d, t)]$ of numbers of copies of allele A_i ($i = 1, 2, \ldots m$) in the female population; $\boldsymbol{p}_i(t)$ is the corresponding vector of frequencies of A_i across female age-classes. We have

$$\boldsymbol{n}_i(t) = \sum_{j=1}^{m} [\boldsymbol{n}_{ij}(t) + \boldsymbol{n}_{ji}(t)] \qquad (2.1a)$$

$$p_i(t) = \tfrac{1}{2} \sum_{j=1}^{m} [p_{ij}(t) + p_{ji}(t)] \tag{2.1b}$$

Obviously, similar vectors n_i^* and p_i^* can be defined for the male population.

It turns out that, much as in the usual discrete-generation case (Crow and Kimura, 1970, Chapter 2), it is most convenient to study the process of genetic change in terms of gene frequencies or gene numbers rather than genotypic frequencies. We shall see below that, since each genotype is assumed to have the same demographic parameters, the gene frequencies must approach constant values which are independent of sex and age. Once these equilibrium gene frequencies have been effectively attained, the equilibrium genotypic frequencies can be calculated knowing the rules of mating, e.g. with random mating.

Our method, therefore, will be to work directly with the vectors n_i and n_i^*. The first problem is to relate the values of these vectors for some arbitrary time t to their values in the preceding time-interval, $t-1$. This can be done by means of a matrix equation similar to that employed in Chapter 1 (section 1.2.1) for describing the dynamics of a genetically homogeneous population. We use the same notation as in that chapter; thus $P(x, t)$ and $P^*(x, t)$ are the probabilities of survival over one time-interval of a female and male, respectively, who are aged x at time t. The net fecundities $f(x, t)$, $f^*(x, t)$ and $f'(x, t)$, as defined by equations (1.3), (1.8) and (1.10), are also needed. These are, respectively, the expected numbers of daughters of a female aged x at time t, the expected number of sons of a male aged x at time t, and the expected number of sons of a female aged x at time t. In addition, we define $f''(x, t)$ as the expected number of daughters of a male aged x at time t. In the notation of Chapter 1, we have

$$f''(x, t) = M^*(x, t)P(0, t)a^*(x, t) \tag{2.2}$$

As discussed in Chapter 1 (pp. 6–8), the age-specific fecundities of males are in general dependent on the age-composition of the population in a highly complex way, which is determined by the precise rules of mating with respect to the ages of male and female partners. As will be seen shortly, the exact nature of this dependence is irrelevant; all that we need to know is that each genotype at the locus in question has the same demographic parameters.

We can combine n_i and n_i^* into a single $(d + d^*)$-dimensional row vector $\tilde{n}_i = [n_i, n_i^*]$, which describes the state of the population with respect to allele A_i (cf. section 1.2.1). Using the rules of Mendelian transmission, we can write the following matrix equation

$$\tilde{n}_i(t) = \tilde{n}_i(t-1)H(t-1) \qquad (2.3a)$$

The matrix $H(t-1)$ in equation $(2.3a)$ can be written in the following partitioned form

$$H(t-1) = \left[\begin{array}{c|c} A(t-1) & B(t-1) \\ \hline C(t-1) & D(t-1) \end{array} \right] \qquad (2.3b)$$

where A is a $d \times d$ matrix whose only non-zero elements are those of the first column, $a_{x1} = \frac{1}{2}f(x, t-1)$, and the off-diagonal elements, $a_{x,x+1} = P(x, t-1)$; B is a $d \times d^*$ matrix whose only non-zero elements are those of the first column, $b_{x1} = \frac{1}{2}f'(x, t-1)$; C is a $d^* \times d$ matrix whose only non-zero elements are those of the first column, $c_{x1} = \frac{1}{2}f''(x, t-1)$; D is a $d^* \times d^*$ matrix whose only non-zero elements are those of the first column, $d_{x1} = \frac{1}{2}f^*(x, t-1)$, and the off-diagonal elements, $d_{x,x+1} = P^*(x, t-1)$. H is therefore a non-negative matrix. The sub-matrices of H describe the transmission of genes between different classes of individuals, so that H may be regarded as having the following schematic structure

$$\left[\begin{array}{cc} \text{females to females} & \text{females to males} \\ \text{males to females} & \text{males to males} \end{array} \right]$$

The same type of equation can be written for each allele in the population, and hence for the total number of alleles. The elements of the matrix are, from the assumptions which we have made, identical for every allele; only the vectors $\tilde{n}_i(t)$ differ. We show below that this means that every allele in the system comes to increase in numbers at the same rate (which in general, of course, varies in time), so that the frequencies of all the alleles eventually become constant. To do this, we make use of the weak ergodicity property of non-negative matrices, which we discussed in section 1.4.1. As we saw there, the relative values of the components of a vector representing the state of a system described by an equation such as (2.3*a*) eventually become independent of the initial state of the population. Two vectors representing two systems (two alleles) which differ only in their initial vectors become strictly proportional to each other. Applied to the present case, this means that the number of copies of a given allele, A_i, in a given age-class becomes proportional to the number of copies of a different allele, A_j, in that age-class, so that their relative frequencies become constant, but in general depend on the initial conditions.

Analysis of the model. We now develop a proof of the above statements, using the form of the weak ergodicity property derived by Seneta (1973, pp. 69–77). Equation (2.3*a*) can be used to express $\tilde{n}_i(t)$ in terms of the initial vector $\tilde{n}_i(0)$. We have

$$\tilde{n}_i(t) = \tilde{n}_i(0)H^*(t) \tag{2.4a}$$

where

$$H^*(t) = H(0)H(1)\ldots H(t-1) \tag{2.4b}$$

Seneta has shown that a matrix such as H^*, which is formed by taking the product of a sequence of non-negative matrices with bounded components, tends with increasing t to a state in which its rows are proportional to each other. It is assumed in Seneta's theorem that the matrices whose product composes H^* have a structure such that H^* has no zero elements (i.e., it is *strictly positive*) when t is sufficiently large. We assume for the moment that this condition is satisfied. We therefore obtain the following type of asymptotic relation between any two elements which belong to the xth column of $H^*(t)$

$$h_{ux}^*(t)/h_{vx}^*(t) \sim C_{uv} \tag{2.5}$$

C_{uv} is a constant which is independent of x and t, and depends only on u and v (for a given sequence of matrices composing $H^*(t)$).

Given this result, we can write the following asymptotic expression for the number of copies of A_i in age-class x of the female population at time t

$$n_i(x, t) \sim \sum_{u=1}^{d} n_i(u, 0)h_{ux}^*(t) + \sum_{u=1}^{d^*} n_i^*(u, 0)h_{d+u,x}^*(t)$$

or

$$n_i(x, t) \sim h^*_{1x}(t)C'_i \tag{2.6a}$$

where

$$C'_i = \sum_{u=1}^{d} n_i(u, 0)C_{u1} + \sum_{u=1}^{d^*} n^*_i(u, 0)C_{d+u,1} \tag{2.6b}$$

(The C_{u1} are obtained by using equation (2.5) with $v = 1$.) A similar result can be written for the number of copies of A_i in age-class x of the male population

$$n^*_i(x, t) \sim h^*_{1,d+x}(t)C'_i \tag{2.6c}$$

The same type of relationship holds for the number of copies of A_i in any other age-class of males or females. The same constant, C'_i, appears in all these expressions. C'_i is determined by the initial vector $\tilde{n}_i(0)$, and by the relations between the rows of $\boldsymbol{H}^*$ as expressed by equation (2.5). The matrix-derived terms, of the form $h^*_{1.}(t)$, are of course different for different age-classes.

The same argument can obviously be applied to each allele in the population, and expressions similar to equations (2.6) can be written down by simply changing the allele subscript i. The equations for different alleles thus differ only with respect to the constants C'_i ($i = 1, 2, \ldots m$), given by equations of the same form as (2.6b). We have the following asymptotic expression for the frequency of A_i among females of age-class x

$$p_i(x, t) \sim h^*_{1x}(t)C'_i \Big/ h^*_{1x}(t) \sum_{j=1}^{m} C'_j$$

$$= C'_i \Big/ \sum_{j=1}^{m} C'_j = \hat{p}_i \tag{2.7}$$

The same asymptotic frequency, $\hat{p}_i$, clearly holds for each age-class of both the male and female parts of the population, since the $h^*_{1.}(t)$ terms cancel top and bottom.

Hardy–Weinberg equilibrium.   The asymptotic allele frequencies, $\hat{p}_i$, ($i = 1, 2, \ldots m$) are therefore constant, and independent of sex and age-class. This is the desired result. It can be seen from equations (2.5)–(2.7) that any allele which is present initially in at least one reproductive or pre-reproductive age-class of the male or female population must be represented in the final population (note that all the elements of H^* are positive). Furthermore, the gene frequencies will in general converge only gradually towards their asymptotic values; only if the frequencies of each allele are initially the same in each age-class of both the male and female populations will there be no change in gene frequencies over the course of time. This gradual convergence of gene frequencies towards constancy is in contrast to the standard result of population genetics for the corresponding discrete-generation model, where gene frequencies are always constant in the absence of disturbing evolutionary forces (Crow and Kimura, 1970, Chapter 2). If the population mates at random with respect to genotype, it follows from the constancy of the asymptotic allele frequencies that the genotypic frequencies for each sex and age-class are also asymptotically constant, and given by the Hardy–Weinberg formula

$$p_{ij}(x, t) = p_{ij}^*(x, t) = \hat{p}_i \hat{p}_j \tag{2.8}$$

A random-mating population thus converges to a state of Hardy–Weinberg equilibrium.

Positivity of H^*. These results are based on the assumption that $H^*(t)$ is a strictly positive matrix for sufficiently large t (p. 75). We now consider the conditions for the validity of this assumption. It is known that such a result holds for the product of a sequence of Leslie matrices, under conditions discussed in section 1.4.1. Now the sub-matrices A and D in equation (2.3b) are both Leslie matrices, describing the female and male parts of the population respectively. It is reasonable to assume that they meet these conditions if we are describing an iteroparous species (the ages with non-zero fecundities may, of course, be different in males and females). The other sub-matrices contribute additional positive elements. Provided that b_{d1} is always positive, the matrix H in equations (2.3) thus has the form of a Leslie matrix with some additional positive elements. Hence, $H^*(t)$ for sufficiently large t must be strictly positive.

Conclusions. These results establish that the conditions for the approach to Hardy–Weinberg equilibrium in an age-structured population with discrete age-classes are similar to those required for the age-structure of the population to become independent of its initial state (weak ergodicity). These conditions are likely to be satisfied for most iteroparous species. This conclusion applies to populations exposed to environmentally-induced fluctuations in their demographic parameters, as well as to populations with time-independent parameters. Furthermore, since we do not need to specify the functional form of the dependence of the demographic parameters on time to apply this result, it holds good for density-dependent populations, despite the fact that the value of the matrix *H* in equations (2.3) may depend partly on the initial vectors of the total numbers of males and females. The rate of convergence to Hardy–Weinberg equilibrium must be of the same order as the rate of convergence of the population to its asymptotic state of independence of the initial conditions, since the equations obeyed by individual alleles are identical in form to those obeyed by the total number of alleles or the total number of individuals. As we saw in section 1.3.2, the period of time required for this convergence is usually a matter of a few 'generations'. During the process of convergence, oscillations in gene frequencies may occur. This conclusion has been confirmed by computer calculations of population trajectories. An example is shown in Table 2.1.

The possibility of gradual convergence of gene frequencies to constancy, and of genotypic frequencies to Hardy–Weinberg proportions, implies that populations which have recently been founded by a group of individuals in which gene frequencies differed according to age and sex may show significant changes in their genotypic composition in the short-term, over and above any effects attributable to other evolutionary factors such as selection or genetic drift. This may be of significance in relation to studies of the genetics of human isolates, e.g. Jacquard (1974, Chapter 16).

The convergence of the population to a state of identical, constant genotypic frequencies in each age-class is dependent on the matrix *H* having a non-periodic structure such that *H** becomes strictly positive (pp. 75–7). The most biologically important case in which this condition is not fulfilled is that of a semelparous species

Table 2.1. *Convergence to equilibrium for an autosomal locus with two alleles, using demographic data for the French 1830 population. The frequency of A_1 was 1 in the first 5 age-classes and 0 in the others.*

Time t	Age-class	Age-structure[a]	Genotype Frequency (A_1A_1)	(A_1A_2)	(A_2A_2)	Frequency of A_1
2	1	161.9	0.273	0.499	0.299	0.523
	5	0.573	1	0	0	1
	9	0.558	0	0	1	0
8	1	153.1	0.362	0.479	0.158	0.602
	5	0.620	0.842	0.151	0.007	0.918
	9	0.335	1	0	0	1
20	1	153.3	0.513	0.406	0.080	0.717
	5	0.650	0.519	0.403	0.078	0.721
	9	0.504	0.625	0.332	0.044	0.790
40	1	162.9	0.530	0.396	0.074	0.728
	5	0.638	0.531	0.395	0.074	0.729
	9	0.515	0.536	0.392	0.072	0.732
60	1	172.6	0.531	0.395	0.074	0.729
	5	0.637	0.531	0.395	0.074	0.729
	9	0.516	0.531	0.395	0.074	0.729
80	1	182.7	0.531	0.395	0.074	0.729
	5	0.638	0.531	0.395	0.074	0.729
	9	0.516	0.531	0.395	0.074	0.729

[a] The total number of individuals entering age-class 1 at time t is given for age 1, and the numbers entering age-classes 5 and 9 are given as fractions of this number. (After Table 7.4 of Jacquard (1974).)

in which the age of reproduction is at 2 or more (cf. Chapter 1, p. 26). In such a case, the different phases of the population form two or more effectively genetically isolated populations coexisting in the same locality; gene frequencies within each phase will be constant, but any initial differences will be preserved indefinitely. A probable example of such a situation has been described in the pink salmon by Aspinwall (1974).

2.2.2 *An autosomal locus: the continuous-time case*

It is possible to approximate the continuous-time case arbitrarily closely by using the above discrete age-class model with a large number of age-classes. It is instructive, however, to develop an exact continuous-time model incorporating genetic variation and to use it to establish convergence of a random-mating population to a state of Hardy–Weinberg equilibrium. We shall employ the notation and concepts introduced in section 1.2.2 for the purpose of describing the dynamics of a continuous-time population. We shall assume that the primary sex-ratio, a, is independent of age and time. The reproductive function for females aged x at time t is therefore equal to

$$k(x, t) = l(x, t)M(x, t)a \qquad (2.9a)$$

and the reproductive function for males is

$$k^*(x, t) = l^*(x, t)M^*(x, t)(1 - a) \qquad (2.9b)$$

These give, respectively, the expected rate of production of female offspring at time t attributable to a female zygote formed at time $t - x$, and the expected rate of production of male offspring at time t attributable to a male zygote formed at time $t - x$. The corresponding expected rates of production of offspring of both sexes are

$$K(x, t) = k(x, t)/a \qquad (2.10a)$$

and

$$K^*(x, t) = k^*(x, t)/(1 - a) \qquad (2.10b)$$

As in the discrete age-class model discussed above, the male fecundities are dependent on the composition of the female population in a complicated way. It is unnecessary for the present purpose to concern ourselves with the nature of this dependence; as previously, all that we need is the fact that all the demographic parameters are the same for each genotype.

It is most convenient to describe the state of the population, as far as total numbers of individuals are concerned, in terms of the rates of production of female and male zygotes by the population at time t, $B_f(t)$ and $B_m(t)$, respectively. From the definitions given above and in Chapter 1, we have

$$B_f(t) = g(t) + \int_0^t B_f(t - x)k(x, t)\, dx \qquad (2.11a)$$

$$B_m(t) = g^*(t) + \int_0^t B_m(t - x)k^*(x, t)\, dt \qquad (2.11b)$$

where $g(t)$ and $g^*(t)$ are terms representing the contributions to $B_f(t)$ and $B_m(t)$ from individuals alive at the initial time $t = 0$; these terms are zero for $t > d$ and $t > d^*$, respectively. Using the assumption of constant sex-ratio a, it is easy to see that the rate of production of zygotes of both sexes, $B(t)$, is related to $B_f(t)$ and $B_m(t)$ by the expressions

$$B_f(t) = aB(t) \tag{2.12a}$$

$$B_m(t) = (1-a)B(t) \tag{2.12b}$$

Using these relations and equations (2.11), we obtain the following expressions for $B(t)$

$$B(t) = \frac{1}{a} g(t) + a \int_0^t B(t-x)K(x, t)\, dx \tag{2.13a}$$

$$B(t) = \frac{1}{(1-a)} g^*(t) + (1-a) \int_0^t B(t-x)K^*(x, t)\, dx \tag{2.13b}$$

Substituting from equations (2.9) and (2.10) and adding the resulting expressions, we obtain

$$B(t) = G(t) + \tfrac{1}{2} \int_0^t B(t-x)[k(x, t) + k^*(x, t)]\, dx \tag{2.14}$$

where $G(t) = \tfrac{1}{2}\{(1/a)g(t) + [1/(1-a)]g^*(t)\}$. This is a renewal equation of the same general form as that which Norton (1928) showed has a weak ergodicity property, under suitable conditions. Two such equations, which differ only in the values of the initial conditions, $G(t)$, tend asymptotically to a state in which the $B(t)$ function for one has a constant ratio to the $B(t)$ function for the other (cf. section 1.4.1). Sufficient conditions for this result to be valid are that $k(x, t)$ and $k^*(x, t)$ be continuous functions of x and t, and that the ages of first and last reproduction for males and females, b, b^*, d and d^*, be fixed over time. If it can be shown that each *allele* in the population satisfies an equation of similar form to equation (2.14), differing only in the initial conditions, then Norton's theorem can be applied to prove that allele frequencies tend to constancy as t increases. We shall now proceed to demonstrate that this is indeed the case.

Since each individual receives one copy of an autosomal gene at a locus from its male parent and the other from its female parent, it is essential to distinguish between alleles of maternal and paternal origin. Consider the zygotes produced by the population at the instant t. Let $p_i(t)$ and $p_i^*(t)$ be the frequencies of a given allele A_i among maternally-derived and paternally-derived alleles, respectively, in these zygotes. The rates of production at time t of maternally-derived and paternally-derived copies of A_i in new zygotes are thus given by $B(t)p_i(t)$ and $B(t)p_i^*(t)$. (Since we are dealing with the case of an autosomal locus, it is unnecessary to distinguish the sexes when calculating gene frequencies among new zygotes.) The net frequency of A_i among the zygotes produced at time t is given by

$$\bar{p}_i(t) = \tfrac{1}{2}[p_i(t) + p_i^*(t)] \tag{2.15a}$$

and the net rate of production of A_i genes in new zygotes is

$$2B(t)\bar{p}_i(t) = B(t)[p_i(t) + p_i^*(t)] \tag{2.15b}$$

Using these definitions, it is possible to derive a renewal equation for $B(t)\bar{p}_i(t)$ in a way similar to that which yielded equation (2.14).

Consider first a time $t > d$, d^*, such that all new zygotes entering the population are derived from individuals who were born after the initial time $t = 0$. Using equations (2.9), (2.10) and (2.13), we obtain

$$B(t)p_i(t) = \int_b^d B_f(t-x)\bar{p}_i(t-x)K(x,t)\,\mathrm{d}x$$

$$= \int_b^d B(t-x)\bar{p}_i(t-x)k(x,t)\,\mathrm{d}x \tag{2.16a}$$

Similarly,

$$B(t)p_i^*(t) = \int_{b^*}^{d^*} B_m(t-x)\bar{p}_i(t-x)K^*(x,t)\,\mathrm{d}x$$

$$= \int_{b^*}^{d^*} B(t-x)\bar{p}_i(t-x)k^*(x,t)\,\mathrm{d}x \tag{2.16b}$$

For $t \le d$, equation (2.16a) must be modified by addition of a term $g_i(t)$, to take account of contributions from individuals alive initially. Similarly, for $t \le d^*$, equation (2.16b) must be modified by the addition of a term $g_i^*(t)$. The exact form of these terms does not concern us here. If we write $G_i(t) = \frac{1}{2}[g_i(t) + g_i^*(t)]$, equations (2.16) can be combined to yield the final equation

$$B(t)\bar{p}_i(t) = G_i(t) + \frac{1}{2}\int_0^t B(t-x)\bar{p}_i(t-x)[k(x, t) + k^*(x, t)]\,dx \quad (2.17)$$

This, as anticipated above, has exactly the same form for each allele, which is that of equation (2.14) for the total number of individuals in the population. Only the nature of the initial conditions, G_i, differ between alleles. Hence, Norton's theorem can be applied, and we can conclude that asymptotically the allele frequencies tend to constancy, i.e. for large t we have

$$\bar{p}_i(t) = \hat{p}_i \quad (i = 1, 2, \ldots m) \tag{2.18}$$

where the values of the equilibrium frequencies $\hat{p}_i$ depend on the initial conditions G_i. These frequencies are clearly the same for males and females, from what was said above; similarly, for large t, such that gene frequency change among the zygotes has ceased, the gene frequencies among individuals of any age must also be constant and identical to the frequencies among the zygotes. When this state has been reached, it also follows that there is Hardy–Weinberg equilibrium for the genotypic frequencies among individuals of all ages, assuming that the population mates at random with respect to genotype. This result has been derived under conditions on the time-dependence of the demographic parameters which are almost as non-restrictive as for the discrete age-class model; the assumption of a constant primary sex-ratio has had to be made, however, in order to obtain equations to which Norton's weak ergodicity theorem can be applied. This is not an unduly restrictive condition, since empirical evidence suggests that there are only minor variations in sex-ratio with time and age of parent (cf. section 5.4.3).

2.2.3 A sex-linked locus

This case can be treated in a very similar way to an autosomal locus, taking into account the fact that a male (assumed to be the heterogametic sex) receives no genes from his father, whereas females receive one gene from each parent. Only the discrete age-class model will be considered here. A special case has previously been treated by Cornette (1978).

As in the autosomal case, an abbreviated description of the state of the population is given by the vectors of numbers of copies of each allele in females and males, $n_i(t)$ and $n_i^*(t)$ $(i = 1, 2, \ldots m)$. As before, n_i and n_i^* can be combined into one $(d + d^*)$-dimensional vector $\tilde{n}_i = [n_i, n_i^*]$ for the purposes of calculation. We can write

$$\tilde{n}_i(t) = \tilde{n}_i(t-1)\tilde{H}(t-1) \qquad (2.19a)$$

where

$$\tilde{H}(t-1) = \left[\begin{array}{c|c} \tilde{A}(t-1) & \tilde{B}(t-1) \\ \hline \tilde{C}(t-1) & \tilde{D}(t-1) \end{array}\right] \qquad (2.19b)$$

in which $\tilde{A}$ is a $d \times d$ matrix whose only non-zero elements are those of the first column, $\tilde{a}_{x1} = \frac{1}{2}f(x, t-1)$, and the off-diagonal elements $\tilde{a}_{x,x+1} = P(x, t-1)$; $\tilde{B}$ is a $d \times d^*$ matrix whose only non-zero elements are those of the first column, $\tilde{b}_{x1} = \frac{1}{2}f'(x, t-1)$; $\tilde{C}$ is a $d^* \times d$ matrix whose only non-zero elements are those of the first column, $\tilde{c}_{x1} = f''(x, t-1)$; $\tilde{D}$ is a $d^* \times d^*$ matrix whose only non-zero elements are those of the off-diagonal, $\tilde{d}_{x,x+1} = P^*(x, t-1)$.

$\tilde{H}$ is of a somewhat different form from H, the matrix for the autosomal case. As in that case, however, it is equivalent to a Leslie matrix with some additional positive elements, provided that $\tilde{b}_{d1}$ is positive. The same type of argument that was used for an autosomal locus can therefore be applied to this case to prove that the frequencies of A_i among males and females of each age-class approach the same constant value $\hat{p}_i$, at about the same rate as the age-structure of the whole population approaches a state in which it is independent of the initial conditions. The conditions under which this is true are the same as for an autosomal locus. Since the allele frequencies in males and females become the same, the frequency of the ordered genotype A_iA_j in females is given by an equation of the same form as equation (2.8). The genotypic frequencies in males are, of course, identical with the corresponding allele frequencies. These results are similar to those for the standard discrete-generation model of a sex-linked locus in a random-mating population

(Crow and Kimura, 1970, Chapter 2). There may be some difference in the mode of approach to equilibrium, however. In the discrete generation case, there is a damped oscillation if male and female gene frequencies are initially different. With age-structure, such an oscillation would tend to be disguised by differences between age-classes; only if *all* age-classes of males in the initial population differed in gene frequencies from *all* age-classes of females in the same direction, would one expect oscillations as marked as in the discrete generation case (cf. Cornette, 1978).

2.2.4   *Two autosomal loci*

We shall now consider the case of two autosomal loci. In the discrete-generation case, it is known that the gamete frequencies for such a pair of loci eventually approach values corresponding to random combinations of alleles at the two loci, i.e. the frequency of a particular gamete in the population approaches asymptotically the product of the frequencies of the alleles which it contains. This was first proved by Robbins (1918). A similar result can be proved for the case of an age-structured population, by means of a generalisation of a method introduced by Malécot (1948) for discrete-generation populations. An account of Malécot's results is given by Crow and Kimura (1970, Chapter 2). Random mating with respect to genotype is assumed.

The problem is most conveniently treated by means of a difference equation formulation of the discrete age-class model (cf. Chapter 1, pp. 10–12. The continuous-time case follows from this straightforwardly, by increasing the number of age-classes without limit. Consider two autosomal loci A and B, with alleles A_i ($i = 1, 2, \ldots m$) and B_k ($k = 1, 2, \ldots n$). The genetic state of the system in a given time-interval t can be characterised by the frequencies of the nm possible gametes, A_1B_1, A_1B_2, $\ldots A_mB_n$, among the zygotes born in time-interval t. Let $\pi_{ik}(t)$ be the frequency of gametes containing alleles A_i and B_k among the gametes of maternal origin in these zygotes. Let the corresponding frequency for paternally-derived gametes be $\pi_{ik}^*(t)$. The overall frequency of A_iB_k gametes in the zygotes is thus given by

$$\bar{\pi}_{ik}(t) = \tfrac{1}{2}[\pi_{ik}(t) + \pi_{ik}^*(t)] \tag{2.20a}$$

As we have seen, the frequencies of alleles at autosomal loci tend to constancy with increasing time, under the conditions discussed in section 2.2.1. Let the equilibrium frequencies be $\hat{p}_i$ for A_i and $\hat{q}_k$ for B_k. We shall assume from now on that the population has effectively reached such a state of constant gene frequencies. Since we are concerned here only with establishing that gamete frequencies approach those predicted from random combination of alleles, there is no loss in generality in making this assumption. If the alleles were combined at random into gametes, the frequency of A_iB_k gametes would be equal to $\hat{p}_i\hat{q}_k$. The deviation from random combination can be measured by the *linkage disequilibrium* parameter D_{ik}, defined by the equation

$$D_{ik}(t) = \bar{\pi}_{ik}(t) - \hat{p}_i\hat{q}_k \qquad (2.20b)$$

We shall show that this parameter approaches zero with increasing t.

We proceed in a somewhat similar manner to the analysis of the case of a single autosomal locus with continuous time (section 2.2.2); in particular, we make the assumption that the primary sex-ratio a is constant, so that the total number of zygotes born at time t, $B(t)$, is related to the numbers of male and female zygotes, $B_m(t)$ and $B_f(t)$, by equations of the same form as equations (2.12). This enables us to write the following equation for $B(t)$ (assuming $t \geqq d, d^*$, so that no zygotes are contributed by individuals alive at the initial time $t = 0$)

$$B(t) = \sum_{x=b}^{d} B(t-x)k(x, t) = \sum_{x=b^*}^{d^*} B(t-x)k^*(x, t) \qquad (2.21)$$

Let the *recombination fraction* for loci A and B be c in females and c^* in males. For simplicity, we shall assume that these are independent of age, although it is not difficult to generalise the following argument to allow for age-dependence of recombination fractions. Consider the set of gametes produced by females aged x at a time $t > d, d^*$. There is a probability $1 - c$ that a gamete drawn at random from this set is non-recombinant, in which case there is a probability $\bar{\pi}_{ik}(t-x)$ that it is A_iB_k in constitution. There is a probability c that it is recombinant in origin, in which case there is a probability $\hat{p}_i\hat{q}_k$ of its being A_iB_k. The net frequency of A_iB_k among these gametes is therefore equal to

$$(1-c)\bar{\pi}_{ik}(t-x)+c\hat{p}_i\hat{q}_k$$

Summing over all ages, and using the assumption of constant sex-ratio, we thus obtain the relation

$$B(t)\pi_{ik}(t)=\sum_{x=b}^{d} B(t-x)[(1-c)\bar{\pi}_{ik}(t-x)+c\hat{p}_i\hat{q}_k]k(x,t)$$

$$(2.22a)$$

A similar argument can be applied to paternally derived gametes, and yields the equation

$$B(t)\pi_{ik}^*(t)=\sum_{x=b^*}^{d^*} B(t-x)[(1-c^*)\bar{\pi}_{ik}(t-x)+c^*\hat{p}_i\hat{q}_k]k^*(x,t)$$

$$(2.22b)$$

By using equation (2.21), subtracting $B(t)\hat{p}_i\hat{q}_k$ from both sides of equations (2.22), and adding the resulting equations, we obtain the following expression

$$B(t)D_{ik}(t)=\tfrac{1}{2}(1-c)\sum_{x=b}^{d} B(t-x)D_{ik}(t-x)k(x,t)$$

$$+\tfrac{1}{2}(1-c^*)\sum_{x=b^*}^{d^*} B(t-x)D_{ik}(t-x)k^*(x,t) \qquad (2.23)$$

Choose a value of t, t_0 say, which can be made arbitrarily large. Let $\bar{b}$ be the smaller of b and b^*, and $\bar{d}$ be the larger of d and d^*. Let $|D_{ik}(t_1)|$ be the *maximum* value of $|D_{ik}(t)|$ in the interval $[t_0-\bar{d}, t_0-\bar{b}]$. Equation (2.23) gives the inequality

$$B(t_0)|D_{ik}(t_0)|\leq\tfrac{1}{2}(1-c)|D_{ik}(t_1)|\sum_{x=b}^{d} B(t_0-x)k(x,t_0)$$

$$+\tfrac{1}{2}(1-c^*)|D_{ik}(t_1)|\sum_{x=b^*}^{d^*} B(t_0-x)k^*(x,t_0)$$

Using equations (2.21), this reduces to

$$|D_{ik}(t_0)|\leq[1-\tfrac{1}{2}(c+c^*)]|D_{ik}(t_1)| \qquad (2.24)$$

$|D_{ik}(t_1)|$ can similarly be related to the maximum vaue of $|D_{ik}(t)|$, $|D_{ik}(t_2)|$, in the interval $[t_1-\bar{d}, t_1-\bar{b}]$, and so on. Continuing this back l steps to some arbitrary time t_l $(t_l \geqq \bar{d})$, we have

$$|D_{ik}(t_0)| \leqq [1 - \tfrac{1}{2}(c + c^*)]^l |D_{ik}(t_l)| \tag{2.25}$$

If t_0 is taken sufficiently large, $|D_{ik}(t_0)|$ can be made arbitrarily small, if c or $c^* > 0$, since $|D_{ik}(t_l)|$ is bounded (its maximum possible value is $\tfrac{1}{4}$), and $\tfrac{1}{2} \leqq 1 - \tfrac{1}{2}(c + c^*) < 1$.

This proves that the frequency of A_iB_k averaged over maternal and paternal gametes, $\bar{\pi}_{ik}$, approaches a constant value, $\hat{p}_i\hat{q}_k$, which corresponds to random combination of the alleles concerned. From equations (2.22) it follows that the frequencies of A_iB_k in maternal and paternal gametes, π_{ik} and π_{ik}^*, also approach this limiting value. The same obviously applies to each gamete in the population. We may expect the average length of the time-interval between successive t_i to be of the same order as one of the generation time measures introduced in section 1.3.2; hence the rate of decay of the D parameters given by equation (2.25) is such that each D is reduced by a factor of approximately $1 - \tfrac{1}{2}(c + c^*)$ every 'generation'. This corresponds to the standard result for discrete-generation populations (Crow and Kimura, 1970, Chapter 2).

2.3   The effects of finite population size

2.3.1   *General considerations*

In this section we shall study the process of random changes in gene frequencies due to finite population size (*genetic drift*), in the context of an age-structured population. We assume that other evolutionary forces, such as selection, mutation, migration and non-random mating, are absent. The only factor tending to perturb gene frequencies from the values which they would attain in an infinitely large population is the effect of random sampling of alleles between successive time-intervals. The traditional discrete-generation models of population genetics have dealt with this problem by means either of a measure of the rate of increase in the *inbreeding coefficient* (the probability that the two genes carried at an autosomal locus of a diploid organism are identical by descent), or by examining the rate of increase in the variance of the probability distribution of gene frequencies generated by random drift.

Crow and Kimura (1970, Chapter 7) review this topic, with reference to a number of classes of discrete-generation populations.

In many cases, the asymptotic rates of increase in both these measures are governed by similar equations, at least to a good approximation. If this is so, the rate of genetic drift can be summarised by a single parameter, the *effective population number*, N_e. This concept was introduced by Wright (1931, 1938). For example, let the inbreeding coefficient in generation t be $F(t)$. For sufficiently large t, it is found with discrete-generation models that we can write the approximate equation

$$F(t) \cong 1/2N_e + (1 - 1/2N_e)F(t-1) \qquad (2.26a)$$

Similarly, if there are two alleles A_1 and A_2 with expected frequencies $\hat{p}_1$ and $\hat{p}_2 = 1 - \hat{p}_1$, the variance in the frequency of A_1, $V(t)$, asymptotically satisfies the expression

$$V(t) \cong \frac{\hat{p}_1\hat{p}_2}{2N_e} + \left(1 - \frac{1}{2N_e}\right)V(t-1) \qquad (2.26b)$$

These relations would be satisfied for *all* values of t in a population with the simplest type of structure: a random-mating but self-fertile hermaphrodite species with a constant adult population size N, and a distribution of number of offspring per adult which is binomial with a mean of unity. In such an ideal population, we simply have $N_e = N$. In discrete-generation populations with a more complicated breeding structure, relations such as (2.26) may be obeyed only asymptotically, and N_e may differ from the census number of breeding adults in a way which is calculable from the breeding structure of the population. One difficulty with the use of effective number as a measure of the rate of genetic drift is that the value of N_e in equation (2.26a) is sometimes found to differ from that in equation (2.26b). In such cases, it is necessary to distinguish between *inbreeding effective number* and *variance effective number* (Kimura and Crow, 1963).

We shall accordingly concentrate on trying to define a measure of effective population number for the case of an age-structured population. As mentioned in the introduction to this chapter, this problem has been tackled by a number of authors. The approach used here is based on that of Felsenstein (1971), as modified by

Johnson (1977*b*) and Emigh and Pollak (1979) for the case of a two-sex population. Some results on probabilities of fixation of genes in a finite population will also be given. The final section deals with the problem of calculating the frequencies of consanguineous matings in an isolated population, a question of some importance in human genetics.

2.3.2 *Effective population number with age-structure*
Construction of the model. We shall study in detail only the case of a stationary diploid, discrete age-class population with two sexes and with time-independent demographic parameters for both males and females. Furthermore, we make the somewhat artificial assumption that the numbers of males and females in each age-class are fixed over time. Survival from age-class $x - 1$ to x ($x > 1$) is thus assumed to be random with respect to genotype, but there is interdependence between the deaths of different individuals such that constant numbers $n(x)$ and $n^*(x)$ of females and males are maintained in age-class x. Similarly, although new-born individuals are assumed to be formed by combining gametes sampled at random from infinite pools of male and female gametes, births take place in such a way as to preserve constant numbers of males and females in the first age-class. This model is reasonably realistic for a fairly large population whose size is regulated by density-dependent factors, but is obviously unrealistic for very small populations, whose size and age-structure must themselves be subject to sto-chastic fluctuations. A further assumption which is needed is that mating is at random with respect to age and genotype, i.e. the age and genotype of the father of an individual are independent of the age and genotype of its mother.

We shall mostly be concerned with studying the process of genetic drift in terms of inbreeding. To do this, it is necessary to generalise the concept of inbreeding coefficient in the following way. We introduce a square symmetric matrix of order $d + d^*$, $\boldsymbol{F}(t)$, which describes the state of the system at time t. The xyth element of $\boldsymbol{F}(t)$, $f_{xy}(t)$ ($1 \leqq x \leqq d, 1 \leqq y \leqq d$), is the probability of identity of two genes sampled at random (with replacement) from females of age-classes x and y at time t (for brevity, the term 'identity' will be used here as equivalent to identity by descent). Similarly, $f_{x,d+y}(t)$

$(1 \leqq x \leqq d, 1 \leqq y \leqq d^*)$ is the probability of identity of a gene sampled from a female aged x with one from a male aged y at time t; $f_{d+x,d+y}(t)$ $(1 \leqq x \leqq d^*, 1 \leqq y \leqq d^*)$ is the probability of identity of two genes sampled at random from males aged x and y at time t. In order to avoid clumsiness of notation, whenever it is unnecessary to specify the sex of the individuals involved, the ijth element of $\boldsymbol{F}$ is referred to as the probability of identity of a gene sampled from an individual of 'class' i with a gene sampled from an individual of 'class' j $(1 \leqq i \leqq d + d^*, 1 \leqq j \leqq d + d^*)$. Note that f_{ii} is non-zero even for an outbred population, since two genes sampled from the same class have a chance of being the same.

Analysis of the model. The problem therefore reduces to one of determining a workable expression for $\boldsymbol{F}(t)$ in terms of $\boldsymbol{F}(t-1)$. In order to do this, it is useful to introduce a further matrix, $\boldsymbol{G}$, which describes the flow of genes between different classes of individuals in successive time-intervals in a somewhat different way from the matrix $\boldsymbol{H}$ used earlier (section 2.2.1). In general terms, the ijth element of $\boldsymbol{G}$, g_{ij}, is defined as the probability that a gene of an individual of class j is derived from an individual of class i of the previous time-interval. Matrices of this type were introduced into population genetics by Hill (1972, 1974).

The elements of $\boldsymbol{G}$ can be derived straightforwardly, using the assumptions of constancy of the demographic parameters, population size and age-structure, and the fact that each individual has one maternal and one paternal allele at an autosomal locus. Using the fecundity parameters discussed earlier in this chapter (section 2.2.1), but omitting any dependence on time, we have

$$g_{x1} = \frac{n(x)f(x)}{2 \sum n(y)f(y)} \quad (x = 1, 2, \dots d) \tag{2.27a}$$

$$g_{d+x,1} = \frac{n^*(x)f''(x)}{2 \sum n^*(y)f''(y)} \quad (x = 1, 2, \dots d^*) \tag{2.27b}$$

These define, respectively, the probabilities that a randomly-chosen gene of a female of age-class 1 is derived from a female or male aged x in the preceding time-interval. Similar probabilities can be defined for a gene of a male of age-class 1

$$g_{x,d+1} = \frac{n(x)f'(x)}{2 \sum n(y)f'(y)} \quad (x = 1, 2, \ldots d) \quad (2.27c)$$

$$g_{d+x,d+1} = \frac{n^*(x)f^*(x)}{2 \sum n^*(y)f^*(y)} \quad (x = 1, 2, \ldots d^*) \quad (2.27d)$$

The other elements of G may be found as follows. Clearly, a female of age-class x $(x > 1)$ receives her genes from a female belonging to age-class $x - 1$ in the preceding time-interval with probability 1, and similarly for males. Hence, we have

$$g_{x-1,x} = 1 \quad (x = 1, 2, \ldots d) \quad (2.27e)$$

$$g_{d+x-1,d+x} = 1 \quad (x = 1, 2, \ldots d^*) \quad (2.27f)$$

All other elements of G are zero.

If we assume that the primary sex-ratio is constant and independent of parental age, these expressions can be simplified as follows (cf. Chapter 1, pp. 11–12, 27–8)

$$g_{x1} = g_{x,d+1} = \tfrac{1}{2}k(x) \quad (1 \leq x \leq d) \quad (2.28a)$$

$$g_{d+x,1} = g_{d+x,d+1} = \tfrac{1}{2}k^*(x) \quad (1 \leq x \leq d^*) \quad (2.28b)$$

G is similar in general structure to the H matrix used earlier in the chapter, and may similarly be regarded as being composed of four sub-matrices, each of which describes the flow of genes between a particular sub-set of the population (cf. p. 74). Unlike H, however, it is constant in time, and its elements describe probabilities of transmission, not net numbers of genes transmitted. For future purposes, it is useful to note that G is the transpose of a *stochastic matrix*, since the elements in each column sum to unity. From the standard results concerning stochastic matrices (e.g. Jacquard, 1974, Appendix 2), we can therefore conclude that G has a leading eigenvalue of unity, to which there corresponds a row eigenvector whose elements are all equal. An associated column eigenvector, q^{T}, can be found by the method used for defining reproductive value (equation (1.53)). We have

$$q(x) = \sum_{y=x}^{d} (g_{y1} + g_{y,d+1}) \quad (x = 1, 2, \ldots d) \tag{2.29a}$$

$$q(d+x) = \sum_{y=x}^{d^*} (g_{d+y,1} + g_{d+y,d+1}) \quad (x = 1, 2, \ldots d^*) \tag{2.29b}$$

A measure of the generation time of the population, analogous to the measures defined in section 1.3.2, is given by

$$\tilde{T} = \tfrac{1}{2} \sum_{x=1}^{d+d^*} q(x) \tag{2.30}$$

If primary sex-ratio is constant and independent of parental age, we can use equations (2.28) to obtain the simplified expressions

$$q(x) = \sum_{y=x}^{d} k(y) \quad (x = 1, 2, \ldots d) \tag{2.31a}$$

$$q(d+x) = \sum_{y=x}^{d^*} k^*(y) \quad (x = 1, 2, \ldots d^*) \tag{2.31b}$$

$$\tilde{T} = \tfrac{1}{2} \sum_{x=1}^{d} xk(x) + \tfrac{1}{2} \sum_{x=1}^{d^*} xk^*(x) \tag{2.31c}$$

In this case, $\tilde{T}$ is equal to the mean over the female and male populations of the generation time T defined on p. 33 (note that we are assuming a constant population size).

Using these relations, we can write an asymptotic expression for G^t, using the same methods that yielded equation (1.55a). For large t, we have

$$G^t \sim A = q^{\mathrm{T}} 1 / 2\tilde{T} \tag{2.32}$$

where 1 is a $(d + d^*)$-dimensional row vector, whose elements are all equal to unity. The validity of the asymptotic expression (2.32) is guaranteed if the age-specific fecundities for males and females are both non-periodic, since G then has the structure of a non-periodic Leslie matrix with additional positive elements (cf. equation (1.6)).

We are now in a position to use these definitions to obtain an expression for $F(t)$. It is in practice somewhat more convenient to work with the probabilities $1-f_{ij}(t)$ of *non-identity* between two genes sampled at random from individuals of class i and class j at time t. (If we define $J = \mathbf{1}^T\mathbf{1}$ as the square matrix of order $d+d^*$ whose elements are all unity, we obtain the matrix of these probabilities of non-identity as $J - F(t)$.) Consider firstly $1-f_{11}(t)$, the probability of non-identity between two genes sampled at random from females of age-class 1 at time t. In order for these genes to be non-identical, the same gene must not have been sampled twice from the same female. There is a probability of $1/2n(1)$ that the two genes are separate but come from the same female, in which case there is a probability $4g_{x1}g_{d+y,1}$ that they come from a female aged x and a male aged y at time $t-1$, with probability of non-identity $1-f_{x,d+y}(t-1)$. There is a probability $1-1/n(1)$ that the two genes are sampled from two different females; if this is so, there is a probability $g_{i1}g_{j1}$ that they are derived from individuals belonging to classes i and j at time $t-1$, with probability of non-identity $1-f_{ij}(t-1)$. We thus obtain the recurrence relation

$$1-f_{11}(t) = \frac{2}{n(1)} \sum_{x=1}^{d} \sum_{y=1}^{d^*} g_{x1}g_{d+y,1}[1-f_{x,d+y}(t-1)]$$

$$+\left[1-\frac{1}{n(1)}\right] \sum_{i,j=1}^{d+d^*} g_{i1}g_{j1}[1-f_{ij}(t-1)] \quad (2.33a)$$

A similar expression can be derived for genes sampled from males of age-class 1:

$$1-f_{d+1,d+1}(t)$$

$$= [2/n^*(1)] \sum_{x=1}^{d} \sum_{y=1}^{d^*} g_{x,d+1}g_{d+y,d+1}[1-f_{x,d+y}(t-1)]$$

$$+[1-1/n^*(1)] \sum_{i,j=1}^{d+d^*} g_{i,d+1}g_{j,d+1}[1-f_{ij}(t-1)] \quad (2.33b)$$

An expression for $1-f_{xx}(t)$ ($1 < x \leq d$), the probability of non-identity between two genes sampled from females of age-class x ($x > 1$), can be obtained as follows. The probability that the two genes sampled are separate genes is $1-1/2n(x)$, since the chance that the same gene is sampled twice is $1/2n(x)$. If they are separate genes, they must have been derived from separate genes of females aged $x-1$ at time $t-1$, with probability of non-identity $[1-f_{x-1,x-1}(t-1)]/[1-1/2n(x-1)]$. We thus have

$$1-f_{xx}(t)=[1-f_{x-1,x-1}(t-1)]$$

$$\times\frac{[1-1/2n(x)]}{[1-1/2n(x-1)]} \quad (x=2,3,\ldots d) \quad (2.33c)$$

Similarly,

$$1-f_{d+x,d+x}(t)=[1-f_{d+x-1,d+x-1}(t-1)]$$

$$\times\frac{[1-1/2n^*(x)]}{[1-1/2n^*(x-1)]} \quad (x=2,3,\ldots d^*) \quad (2.33d)$$

For a pair of genes sampled from a male and a female both aged 1 at time t, we have

$$1-f_{1,d+1}(t)=\sum_{i,j=1}^{d+d^*} g_{i1}g_{j,d+1}[1-f_{ij}(t-1)] \quad (2.33e)$$

For a female aged 1 and a male aged x $(x>1)$, we have

$$1-f_{1,d+x}(t)=\sum_{i=1}^{d+d^*} g_{i1}[1-f_{i,d+x-1}(t-1)] \quad (x=2,3,\ldots d^*)$$

$$(2.33f)$$

Similarly, for a female aged x and a male aged 1, we have

$$1-f_{x,d+1}(t)=\sum_{i=1}^{d+d^*} g_{i,d+1}[1-f_{i,x-1}(t-1)] \quad (x=1,2,\ldots d)$$

$$(2.33g)$$

For individuals in all other classes, we have

$$1-f_{ij}(t)=1-f_{i-1,j-1}(t) \quad (2.33h)$$

Equations (2.33) provide us with a set of linear recurrence relations which connect the set of probabilities of non-identity at time t with the set at time $t-1$, in a form which is extremely suitable for numerical calculations. From general theory (cf. section 1.3.1), we expect that asymptotically these probabilities will change with time with a rate determined by the leading eigenvalue, λ_0, of the matrix corresponding to the recurrence relations. The elements of this matrix are easily deducible from equations (2.33). When this asymptotic state has effectively been attained, we can write

$$1 - f_{ij}(t) \sim \lambda_0 [1 - f_{ij}(t-1)]$$

or

$$f_{ij}(t) \sim (1 - \lambda_0) + \lambda_0 f_{ij}(t-1) \quad (i, j = 1, 2, \ldots d + d^*)$$
(2.34a)

Noting that the 'generation time' in the present case is defined by equations (2.30) and (2.31c) as $\tilde{T}$, and comparing equation (2.34a) with equation (2.26a), we can define effective population size by the relation

$$1 - \lambda_0 = 1/2\tilde{T}N_e$$
(2.34b)

$(1 - \lambda_0)\tilde{T}$ predicts the asymptotic per generation increase in the f_{ij}. The problem therefore reduces to one of finding a useful approximate formula for λ_0.

This can be done if we assume that the population size is sufficiently large that second-order terms in $1/n(x)$ and $1/n^*(x)$ can be neglected, and that the asymptotic state represented by equation (2.34a) is approached while the f_{ij} are still small enough that second-order terms in the f_{ij} and terms such as $f_{ij}/n(x)$ are negligible.† If these assumptions are made, we can approximate equations (2.33a and b) for low values of the f_{ij} coefficients by the expressions

$$f_{11}(t) \cong \frac{1}{2n(1)} + \sum_{i,j=1}^{d+d^*} g_{i1} g_{j1} f_{ij}(t-1)$$
(2.35a)

$$f_{d+1,d+1}(t) \cong \frac{1}{2n^*(1)} + \sum_{i,j=1}^{d+d^*} g_{i,d+1} g_{j,d+1} f_{ij}(t-1)$$
(2.35b)

(note that $\sum_{x=1}^{d} g_{x1} = \sum_{y=1}^{d^*} g_{d+y,1} = \frac{1}{2}$, etc.)

---

† The assumption about the rate of convergence to the asymptotic state has been found to be valid whenever computations of specific examples have been carried out (Felsenstein, 1971; Johnson, 1977b; Choy and Weir, 1978). Furthermore, an identical expression for N_e to that obtained here can be derived by the alternative methods of Felsenstein (1971) and Emigh and Pollak (1979). If this assumption breaks down in a particular case, the calculations which follow still provide a valid approximate method for determining the rates of change of the f_{ij} in the early generations of inbreeding.

Similarly, equations (2.33c and d) can be approximated for low f_{ij} by

$$f_{xx}(t) \cong \frac{1}{2n(x)} - \frac{1}{2n(x-1)} + f_{x-1,x-1}(t-1) \quad (x = 2, 3, \dots d)$$

(2.35c)

$$f_{d+x,d+x}(t) \cong \frac{1}{2n^*(x)} - \frac{1}{2n^*(x-1)}$$

$$+ f_{d+x-1,d+x-1}(t-1) \quad (x = 2, 3, \dots d^*)$$ (2.35d)

The remaining members of equations (2.33) can be rewritten in terms of the f_{ij} coefficients alone; in each case a relationship of exactly the same form is preserved merely by substituting f_{ij} for $1 - f_{ij}$ on both sides of the equation. Combining these with equations (2.35), we obtain the single matrix equation

$$\boldsymbol{F}(t) \cong \boldsymbol{G}^T \boldsymbol{F}(t-1)\boldsymbol{G} + \boldsymbol{Y}$$ (2.36)

where $\boldsymbol{Y}$ is a diagonal matrix with non-zero elements $y_{11} = 1/2n(1)$, $y_{xx} = 1/2n(x) - 1/2n(x-1)$ $(x = 2, 3, \dots d)$, $y_{d+1,d+1} = 1/2n^*(1)$, $y_{d+x,d+x} = 1/2n^*(x) - 1/2n^*(x-1)$ $(x = 2, 3, \dots d^*)$. Equation (2.36) gives the relation

$$\boldsymbol{F}(t) \cong (\boldsymbol{G}^T)^t \boldsymbol{F}(0)\boldsymbol{G}^t + \sum_{u=0}^{t-1} (\boldsymbol{G}^T)^u \boldsymbol{Y} \boldsymbol{G}^u$$

where $\boldsymbol{F}(0)$ is the initial value of the $\boldsymbol{F}$ matrix. (If we assume that all individuals in the initial population are unrelated, $\boldsymbol{F}(0)$ is a diagonal matrix with non-zero elements $f_{xx}(0) = 1/2n(x)$ $(x = 1, 2, \dots d)$ and $f_{d+x,d+x}(0) = 1/2n^*(x)$ $(x = 1, 2, \dots d^*)$.)

This gives

$$\boldsymbol{F}(t) - \boldsymbol{F}(t-1) \cong (\boldsymbol{G}^T)^t \boldsymbol{F}(0)\boldsymbol{G}^t - (\boldsymbol{G}^T)^{t-1} \boldsymbol{F}(0)\boldsymbol{G}^{t-1} + (\boldsymbol{G}^T)^{t-1} \boldsymbol{Y} \boldsymbol{G}^{t-1}$$

Substituting from equation (2.32), we obtain

$$\boldsymbol{F}(t) - \boldsymbol{F}(t-1) \cong \boldsymbol{A}^T \boldsymbol{Y} \boldsymbol{A}$$

i.e.

$$\boldsymbol{F}(t) - \boldsymbol{F}(t-1) \cong \frac{(\boldsymbol{q}\boldsymbol{Y}\boldsymbol{q}^T)\boldsymbol{J}}{4\tilde{T}^2}$$ (2.37)

where $\boldsymbol{J}$ is the matrix $\mathbf{1}^T\mathbf{1}$ (p. 94).

The effective population number. Equation (2.37) implies that each f_{ij} coefficient experiences the same increment per time-interval, equal to the scalar $(qYq^T)/(4\tilde{T}^2)$. This can be compared with the expression for the change in f_{ij} given by equations (2.34). Provided that the population size is sufficiently large that equation (2.34a) is a good approximation when the f_{ij} are still small, equations (2.34) can be approximated by

$$f_{ij}(t) - f_{ij}(t-1) \cong \frac{1}{2N_e\tilde{T}} \tag{2.38}$$

A comparison of this with equation (2.37) suggests that the definition of effective population number for the age-structured case should be $1/N_e = (qYq^T)/2\tilde{T}$. Collecting terms in qYq^T, and using equations (2.29), we can write

$$\frac{1}{N_e} = \frac{1}{4\tilde{T}}\left\{\frac{1}{n(1)} + \frac{1}{n^*(1)} + \sum_{x=2}^{d} q^2(x)\left[\frac{1}{n(x)} - \frac{1}{n(x-1)}\right]\right.$$
$$\left. + \sum_{x=2}^{d} q^2(d+x)\left[\frac{1}{n^*(x)} - \frac{1}{n^*(x-1)}\right]\right\} \tag{2.39a}$$

If the primary sex-ratio is independent of parental age and equal to one-half, this expression can be simplified to the following

$$\frac{1}{N_e} = \frac{1}{2\tilde{T}B}\left\{\frac{1}{l(1)} + \frac{1}{l^*(1)} + \sum_{x=2}^{d} q^2(x)\left[\frac{1}{l(x)} - \frac{1}{l(x-1)}\right]\right.$$
$$\left. + \sum_{x=2}^{d^*} q^2(d+x)\left[\frac{1}{l^*(x)} - \frac{1}{l^*(x-1)}\right]\right\} \tag{2.39b}$$

where B is the total number of zygotes produced by the population in each time-interval (cf. Chapter 1, p. 12).

A very similar formula can be derived for the case of an hermaphrodite or monoecious species (Felsenstein, 1971; Johnson, 1977b). If the demographic parameters, etc. are defined as for the female population in the two-sex case, except that fecundities are not corrected for the sex of offspring (cf. Chapter 1, p. 8) we obtain

$$\frac{1}{N_e} = \frac{1}{\tilde{T}B}\left\{\frac{1}{l(1)} + \sum_{x=2}^{d} q^2(x)\left[\frac{1}{l(x)} - \frac{1}{l(x-1)}\right]\right\} \tag{2.40}$$

This is obviously equivalent to equation (2.39b) for the case when male and female demographic parameters are exactly equal.

Equations (2.39) and (2.40) are very useful for displaying the effects of demographic structure on population size. It is immediately obvious that when only individuals of age-class 1 are capable of reproduction, equation (2.39a) reduces to

$$\frac{1}{N_e} = \frac{1}{4N_f} + \frac{1}{4N_m}$$

where N_f and N_m are the numbers of breeding females and breeding males, respectively ($N_f = n(1)$ and $N_m = n^*(1)$). This is of course identical with the standard formula for effective number in the discrete-generation case (Crow and Kimura 1970, Chapter 7). Similarly, for a hermaphrodite species, equation (2.40) gives N_e equal to the census number of breeding adults when only age-class 1 individuals are capable of reproduction. If there are no deaths at all ($l(x) = l^*(x) = 1$ for all x), N_e from equation (2.39b) is equal to $\tilde{T}B$, the total number of individuals born into the population over a period of one generation (note that this assumes a primary sex-ratio of one-half). If there are deaths of individuals only between conception and reaching an age of unity, so that $l(x) = l(1)$ and $l^*(x) = l^*(1)$ for all $x > 1$, then $N_e = 4\tilde{T}/[1/n(1) + 1/n^*(1)]$, i.e., N_e is equal to the harmonic mean of the numbers of females and males in age-class 1 over a period of one generation. It can be seen from equations (2.39) that deaths of individuals in later age-classes reduce N_e below these values, as would be expected intuitively.

Examples of the calculation of effective number.   The use of these formulae for N_e can be illustrated with some of the sets of vital statistics used in Chapter 1 (Tables 1.2 and 1.3). In order to simplify the calculations, it has been assumed that the demographic parameters of males and females are equal, so that equation (2.40) can be used. All the essential features of age-structured populations are preserved in this formula. In order to use the data of Chapter 1, which apply to populations with positive intrinsic rates of increase, for the purpose of calculating N_e from a formula which assumes a stationary population, it is necessary to scale down the age-specific fecundities by dividing them by the net reproduction rate for the population in question, so that the modified age-specific fecundities correspond to those for a stationary population (cf. Felsenstein, 1971). If this is done with the human population data of Table 1.2, we find that

$$N_e = B\tilde{T}/1.03$$

where B is the number of zygotes born into the population in each time-interval, and $\tilde{T}$ is equal to 5.31 time-intervals (a time-interval in this case corresponds to five years). N_e in this case is therefore almost exactly equal to the number of zygotes which enter the population over the course of one generation, which reflects the low mortality of the population in question. It is also interesting to compare N_e with N_b, the total number of breeding adults. N_b can be calculated from B by the formula

$$N_b = B \sum_{x=b}^{d} l(x) \qquad (2.41)$$

In the present case, taking $b = 2$ and $d = 10$, we find $N_b = 8.63B$, so that the effective population number is only 0.60 of the number of individuals in the reproductive age-classes.

A similar calculation can be done for the squirrel data given in Table 1.3. In this case, the length of a time-interval is one year. We find that

$$N_e = B\tilde{T}/7.82$$

where $\tilde{T} = 2.79$. In this case, we have $N_b = 0.60B$, so that $N_e/N_b = 0.59$, a value very similar to that for the human population. The great reduction in N_e below the number of zygotes entering the population per generation is due to the high mortality of the squirrels, particularly in the first year of life.

The inbreeding coefficient of an age-structured population. Up to now we have discussed the problem solely in terms of the matrix of probabilities of identity, $\mathbf{F}$. It is possible, however, to construct a measure similar to the discrete-generation inbreeding coefficient (Johnson, 1977*b*). It is necessary in general to define inbreeding coefficients for males and females separately, although they become equal if the primary sex-ratio is independent of parental age. The inbreeding coefficient for the females at time t, $F(t)$, is defined by analogy with the discrete-generation case as the probability of identity of the maternal and paternal genes of a female in age-class 1. We have

$$F(t) = 4 \sum_{x=1}^{d} \sum_{y=1}^{d^*} g_{x1} g_{d+y,1} f_{x,d+y}(t-1) \qquad (2.42a)$$

Similarly, for males we have the inbreeding coefficient

$$F^*(t) = 4 \sum_{x=1}^{d} \sum_{y=1}^{d^*} g_{x,d+1} g_{d+y,d+1} f_{x,d+y}(t-1) \qquad (2.42b)$$

In practical applications, such as animal breeding problems, it is relatively simple to calculate these inbreeding coefficients for any desired time, by combining equations (2.42) with the results of iterating the exact recurrence relations (2.33), rather than by using the approximate asymptotic equation (2.37), or the estimate of $F(t)$ and $F^*(t)$ of $(t-1)/(2N_e\tilde{T})$ suggested by equations (2.42) and (2.38).

Variance effective number. In certain discrete-generation models, the effective number appropriate for calculating the rate of inbreeding differs from that for the rate of increase in variance in gene frequency (p. 89). It is therefore important to determine whether or not there is any difference in the present case. Johnson (1977*a*) has obtained a formula for variance effective number with age-structure which is identical with equations (2.39); the variance effective number given by Hill (1972, 1979) can also be shown to be the same as that of equations (2.39), for the sampling scheme assumed here (Johnson, 1977*b*). Felsenstein (1971) showed that the inbreeding and variance effective numbers are the same for the present type of model in the case of an hermaphrodite species. His argument can easily be extended to the two-sex case. We can conclude that the variance effective number is identical with the inbreeding effective number in the present case.

Other models. The results described above have been derived on the basis of a number of specialising assumptions, notably that the size and age-structure of the population are constant, and mating is at random with respect to age and genotype. The sampling scheme assumed for the formation of new zygotes implies that each member of a given age-class has an equal chance of contributing to the new-born population (i.e., a binomial distribution of family size within age-classes). Some of these assumptions can be relaxed. Felsenstein (1971) gives a treatment of variance effective number for the case of an hermaphrodite population which is growing in size but has a stable age-structure and constant population growth-rate. Hill (1972, 1979) gives a treatment of variance effective number which is capable of dealing with non-binomial family size distributions (see also Johnson (1977a, b)). Choy and Weir (1978) give recurrence relations for the f_{ij} coefficients which can be used for computations with non-binomial family size distributions. They also give recurrence relationships for digenic identity probabilities.

2.3.3 *The probability of fixation of a gene*

General considerations. In this section, we shall be concerned with the problem of calculating the probability of fixation of a neutral autosomal gene with an arbitrary initial frequency, in an age-structured population of finite size. In the discrete-generation case, the following argument can be used. The formula for the inbreeding coefficient in an arbitrary generation, equation (2.26a), implies that as time increases, F tends to unity, so that all gene copies at a locus are eventually descended with probability unity from one particular gene present in the initial generation. Let the frequency of a given allele A_i be $p_i(0)$ among the females of the initial generation, and $p_i^*(0)$ among the males. With stochastic independence of the offspring distributions for different individuals, the *expected* frequency of A_i in one generation is equal to the actual population gene frequency of the preceding generation, in the absence of mutation, selection, etc. This implies that the expected frequency of A_i in any generation in a finite population is equal to its frequency in an infinite population with the same initial composition, i.e. to $\frac{1}{2}p_i(0) + \frac{1}{2}p_i^*(0)$. Since the final state of the population is such that A_i is either fixed or lost, the probability of fixation of A_i is thus $\frac{1}{2}p_i(0) + \frac{1}{2}p_i^*(0)$.

This type of argument can be extended fairly easily to the case of an age-structured population, using the finite population model developed above. The assumption of random mating with respect to age is, however, unnecessary in this context. The treatment below is based on Emigh and Pollak (1979). (See also Emigh (1979*a*, *b*).)

The calculation of the fixation probability. The definitions of section 2.2.1 can be extended by introducing a d-dimensional vector of expected frequencies of A_i among females of the various age-classes at time t, $\boldsymbol{X}_i(t)$; a corresponding d^*-dimensional vector, $\boldsymbol{X}_i^*(t)$, can be defined for males. As with the gene frequencies themselves, these can be combined into a single $(d+d^*)$-dimensional vector of expected frequencies $\tilde{\boldsymbol{X}}_i(t) = [\boldsymbol{X}_i(t), \boldsymbol{X}_i^*(t)]$. Using the result that, with binomial sampling, the expected frequency of a gene in a given age-class must be equal to the frequency in an infinite population with the same initial state, it is easily shown that

$$\tilde{\boldsymbol{X}}_i(t) = \tilde{\boldsymbol{X}}_i(t-1)\boldsymbol{G} \qquad (2.43a)$$

where $\boldsymbol{G}$ is the matrix of transmission probabilities defined on p. 91. Iterating this equation, we obtain

$$\tilde{\boldsymbol{X}}_i(t) = \tilde{\boldsymbol{X}}_i(0)\boldsymbol{G}^t \qquad (2.43b)$$

where $\tilde{\boldsymbol{X}}_i(0)$ is the initial value of $\boldsymbol{X}(t)$, such that

$$\tilde{\boldsymbol{X}}_i(0) = [\boldsymbol{p}_i(0), \boldsymbol{p}_i^*(0)] \qquad (2.44)$$

Using equation (2.32), equation (2.43b) can be rewritten for large t

$$\tilde{\boldsymbol{X}}_i(t) \sim \tilde{\boldsymbol{X}}_i(0)(\boldsymbol{q}^{\mathrm{T}}\boldsymbol{1})/(2\tilde{T}) \qquad (2.45)$$

The elements of $\tilde{\boldsymbol{X}}_i(t)$ are therefore asymptotically all equal to the same constant value, corresponding to the result derived earlier (section 2.2.1) for gene frequencies in the infinite population model with a general demographic structure. This value is the probability of fixation of A_i, which we can symbolise by U_i. Equation (2.45) yields the expression

$$U_i = \left[\sum_{x=1}^{d} p_i(x, 0)q(x) + \sum_{x=1}^{d^*} p_i^*(x, 0)q(d+x) \right] \Big/ 2\tilde{T} \quad (2.46)$$

A case of particular interest for evolutionary theory is when the allele in question is a new mutant, and is therefore represented initially as a single copy among the zygotes of time $t = 0$. The probability of survival of a new mutant, U, can be found from equation (2.46) as follows. If the mutant individual is female, the gene has an initial frequency of $p_i(1, 0) = 1/2B_f$, where B_f is the total number of female zygotes produced per time-interval. For $x > 1$, we have $p_i(x, 0) = 0$. Hence, for a new mutant among the female part of the population we have

$$U = 1/4B_f\tilde{T} \tag{2.47a}$$

For a new mutant in the male part of the population, we find

$$U = 1/4B_m\tilde{T} \tag{2.47b}$$

where B_m is the number of male zygotes produced per time-interval. These probabilities are equal only when the primary sex-ratio is one-half.

The rate of gene substitution. These results can be used to calculate the rate of substitution of neutral mutations in evolution. The standard method of doing this for a discrete-generation model (Kimura, 1968) assumes that there is an infinite number of possible alleles at a locus, which are selectively equivalent (*neutral alleles*), and that each time a mutation to such an allele occurs in a population, the allele has never previously been represented in the population. If the population has a fixed size N, and the rate of occurrence of neutral mutations at the locus in question is u per generation, then $2Nu$ new mutations appear in the population each generation, on average. But an individual gene present in the population in any particular generation has a probability of $1/2N$ of being fixed by chance. Hence, the expected number of new mutations which enter the population each generation and which eventually displace all the existing alleles at the locus is

$$K = 2Nu/2N = u \tag{2.48}$$

K is the *rate of gene substitution* at the locus in question. If the state of the locus is compared at times t_0 and t_1, where $t_1 - t_0$ is a large number of generations, the expected number of neutral allele substitutions that have occurred is equal to $K(t_1 - t_0)$. This result plays an important role in the interpretation of data on molecular evolution (Kimura and Ohta, 1971; Nei, 1975).

We can extend equation (2.48) to an age-structured population as follows. Consider females of age-class x at a particular time t. A new neutral mutation among these individuals has a frequency of $1/2B_f$, since the mutation was originally present among the female zygotes at time $t - x$. It therefore has a probability of fixation $q(x)/4B_f\tilde{T}$, by equation (2.46). But the expected number of new mutations among the females of age-class x is $2B_f u$, where u is the rate of neutral mutation per time-interval, i.e. the probability that a gene of the zygotes present in a particular time-interval is a newly-mutated neutral allele. It follows that the expected number of new neutral mutations carried by females of age-class x which eventually become fixed is $2B_f u q(x)/4B_f\tilde{T} = uq(x)/2\tilde{T}$. A similar calculation can be done for each age-class of males and females. Summing up over all classes of individuals, we obtain the rate of gene substitution per time-interval as

$$K = \left[u \sum_{x=1}^{d} q(x) + u \sum_{x=1}^{d^*} q(d+x) \right] \Big/ 2\tilde{T}$$

Using equations (2.31), this reduces to

$$K = u \tag{2.49}$$

As in the discrete-generation case, therefore, the rate of gene substitution, under the model of evolution by the chance fixation of neutral mutations, is equal to the neutral mutation rate. The unit of time in equation (2.49) is, of course, the time-interval; the rate for a different time-scale can be found by multiplying by the appropriate factor.

2.3.4 *The frequencies of consanguineous matings in an isolated population*

General considerations. One of the classic problems of human genetics is that of finding the expected frequencies of consanguineous matings in an isolated population. This problem obviously has a close relationship with that of calculating the effective number of a population. It can be stated as follows. Consider an isolated human population of fixed size, and assume that mating is random, so that each individual in the population has the same chance of marrying a given person of the opposite sex. What is the expected frequency of occurrence of consanguineous marriages of a particular degree, e.g. first-cousin marriages? The first attempt to solve this problem was made by Dahlberg (1929, 1948), using a discrete-generation formulation. His approach can be illustrated with the case of first-cousins. Each marriage is assumed to produce two children that survive to maturity, so that the population size is constant. Assuming a sex-ratio of one-half, the number of cousins available to a man of marriageable age can be found as follows. His father had one sibling who married and produced two children. On average, one of these is female, so that the man in question has an expectation of one marriageable female cousin on his father's side. Similarly, he has an expectation of one on his mother's side, giving a total expectation of two marriageable female cousins. If the number of breeding adults is N, there is a total of $\frac{1}{2}N$ girls of marriageable age, so that the probability of the man's marrying a first-cousin is $2/\frac{1}{2}N = 4/N$. If the frequency of first-cousin marriages is known, the population size N can be determined. A similar calculation for uncle–niece and aunt–nephew marriages gives a frequency of $2/N$. These calculations are somewhat artificial, as they assume that there is a fixed family size of two; more realistic models, which allow for a probability distribution of family size, can readily be developed (Frota Pessoa, 1957; Jacquard, 1974, pp. 197–202).

The assumption of discrete generations, which underlies Dahlberg's model and its later generalisations, is inadequate when dealing with data on the human population. The most serious defect of the model is that it leaves out the fact that the probability of marriage is dependent on the ages of the individuals concerned, men being on average older than their wives. The number of potential spouses of a given degree of consanguinity available to a

Figure 2.1. The four kinds of first cousin. (Squares and circles represent males and females respectively)

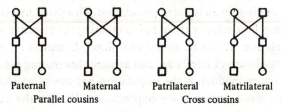

| Paternal | Maternal | Patrilateral | Matrilateral |
| Parallel cousins | | Cross cousins | |

male may therefore differ from the number available to a female. For example, uncle–niece marriages are more common than aunt–nephew marriages, simply because the probability of marriage between a woman and a much older man is higher than the probability of a marriage between a man and a much older woman. On the discrete-generation model, the frequencies of uncle–niece and aunt–nephew marriages should, however, be equal. More subtle effects can arise with marriages of a lower degree of consanguinity. For instance, first-cousin marriages can be divided into four classes, according to the sexes of the ancestors of the partners (Figure 2.1). Data on the frequencies of these classes of marriage in several populations are given in Table 2.2. It can be seen that

Table 2.2. *Observed frequencies of the four kinds of first-cousin marriages* (*relative to a frequency of* 100 *for all first-cousin marriages*)

	Parallel		Cross	
	Paternal	Maternal	Patrilateral	Matrilateral
Germany (1890–1935)	21	29	23	27
Austria (1901–1931)	18	33	21	28
Italy (1851–1957)	22	28	21	29
Japan (1948–1952)	22	33	18	27

After Table 26 of Hajnal (1963)

matrilateral cross-cousin marriages are more frequent than patrilateral cross-cousin marriages, and that maternal parallel marriages are more frequent than paternal parallel marriages. Again, the discrete-generation model predicts equal frequencies of each of these classes. The excess frequency of matrilateral cross-cousin marriages has a similar explanation to that for uncle–niece marriages. Mothers are, on average, younger at the birth of their offspring than fathers, so that a man will tend to have matrilateral cross-cousins who are younger than himself, and hence available for marriage, whereas girls who are his patrilateral cross-cousins will tend to be older than himself. The excess frequency of maternal parallel marriages is harder to explain, and requires an analysis based on the marriage model developed below. This model provides a general solution to the problem of the frequencies of consanguineous mating in an age-structured population, and was formulated by Hajnal (1963, 1976) (see also Cavalli-Sforza, Kimura and Barrai, 1966). The following account is based on that of Hajnal (1963).

Assumptions of the marriage model. We shall assume that the population is closed, so that no individuals enter or leave. Age and time are treated as continuous variables, as is realistic for human populations. The population is assumed to be stationary in size, and to have time-independent demographic parameters. The primary sex-ratio, a, is assumed to be independent of parental age (cf. section 2.2.2). Mating is at random within the population, subject to the restriction that the probability of marriage between a given couple is a known function of their ages. Specifically, we shall assume that the chance of eventual marriage between two individuals selected at birth depends only on the interval between their dates of birth. Let the time of birth of a given boy be subtracted from that of a given girl, and write z for the resulting difference (z is measured in years, and may take negative as well as positive values). The probability that the couple will eventually marry is thus a function of z. This provides a realistic model of the marriage process. Finally, it is assumed that there are no correlations between relatives with respect to their demographic parameters. This is clearly unrealistic, but is the best that can be done in the absence of adequate information on such correlations. Some of these assumptions can be relaxed to a certain extent, as will be discussed below (p. 116).

Consider consanguineous marriages of a given class ξ, specified by a pedigree diagram of the type shown in Figures 2.1 and 2.2. The class ξ is specified by the sequence in which males and females occur in the lines of descent, as well as by the number of steps by which the couple in question trace their descent from a single ancestral couple.† The different categories of first-cousin marriages shown in Figure 2.1 thus correspond to different values of ξ. The problem is to calculate Y_ξ, the average annual number of marriages of class ξ in an isolated population. We can write

$$Y_\xi = AN_\xi\Pi_\xi \tag{2.50}$$

where A is the total number of marriages per year; N_ξ is the expected number of potential marriages of class ξ among the descendants of a couple (i.e. the average number of pairs of males and females related in the manner ξ among the descendants of a couple); Π_ξ is the chance that a potential marriage between such a pair actually takes place.

The calculation of N_ξ. We first examine the relatively straight-forward problem of calculating N_ξ. Consider, for example, the case of paternal parallel cousins. Let a pair of brothers produce during their lifetimes n_1 girl and n_1^* boy babies, and n_2 girl and n_2^* boy babies respectively. The total number of potential marriages among their children is thus $n_1n_2^* + n_1^*n_2$. Taking expectations, and using the above assumption of no correlations between the demographic parameters of relatives, we obtain

$$E[n_1n_2^*] = E[n_1^*n_2] = E[n_1]E[n_2^*] \tag{2.51a}$$

Using the assumptions that the primary sex-ratio, a, is independent of age and that the population is stationary, we have

$$E[n_2^*] = \int_{b^*}^{d^*} k^*(x)\,dx = 1$$

$$E[n_1] = \frac{a}{1-a}\int_{b^*}^{d^*} k^*(x)\,dx = \frac{a}{1-a}$$

Hence

$$E[n_1n_2^*] + E[n_1^*n_2] = \frac{2a}{1-a} \tag{2.51b}$$

---

† Relationships such as half-sibs, which have only one common ancestor, will not be considered here because of the technical difficulties of accounting adequately for illegitimacy in analysing data on marriages involving such relationships (Hajnal, 1963).

If a marriage produces j children, the number of possible pairs of brothers is expected to be $(1-a)^2 j(j-1)/2$, so that the expected number of potential paternal parallel cousin marriages among the descendants of a marriage with j children is, using equation (2.51b), $a(1-a)j(j-1)$. If π_j is the chance that a marriage produces j live births, and if there is no correlation in demographic parameters between parents and offspring, we have

$$N_\xi = a(1-a) \sum_{j=2}^{\infty} j(j-1)\pi_j \qquad (2.52a)$$

The quantity $\sum j(j-1)\pi_j$ depends solely on the mean, μ, and variance, V, of the distribution of numbers of births per marriage. We have

$$\sum_{j=2}^{\infty} j(j-1)\pi_j = \sum_{j=0}^{\infty} j(j-1)\pi_j = V + \mu^2 - \mu$$

Substituting into equation (2.52a), we obtain

$$N_\xi = a(1-a)(V + \mu^2 - \mu) \qquad (2.52b)$$

A similar calculation can be carried out for any class of consanguineous marriage. It is found that N_ξ is independent of ξ, so that equation (2.52b) can be used for any class of marriage.

The calculation of Π_ξ. We now consider the calculation of Π_ξ. Suppose that we select a new-born girl and a new-born boy who have relationship ξ. Let $h_\xi(z)\,\mathrm{d}z$ be the probability that the girl is born between z and $z+\mathrm{d}z$ years after the boy, and $\phi(z)$ be the probability of eventual marriage with a given girl born z years after the boy. We thus have

$$\Pi_\xi = \int_{-\infty}^{\infty} \phi(z)h_\xi(z)\,\mathrm{d}z \qquad (2.53)$$

A formula for $\phi(z)$ can be found as follows. We assume that the population is sufficiently large that stochastic fluctuations in the birth-rate can be neglected. Since the population is stationary, the birth-rate B is constant. Over a short time-interval $\mathrm{d}t$, $aB\,\mathrm{d}t$ girls and $(1-a)B\,\mathrm{d}t$ boys will be born. The numbers of pairs which can be formed among male births occurring between times t and $t+\mathrm{d}t$, and female births occurring between z and $z+dz$ later, is thus $a(1-a)B^2 dt\,dz$. Furthermore, in a stationary population the number of marriages occurring over a short time-interval dt is equal to the number contracted over their entire life-time by the males born in any time-interval of the same length. Let this number be $A\,dt$, where A is the rate of marriage. It follows that

$$A = a(1-a)B^2 \int_{-\infty}^{\infty} \phi(z)\,dz \qquad (2.54a)$$

so that

$$\int_{-\infty}^{\infty} \phi(z)\,dz = A/a(1-a)B^2 \qquad (2.54b)$$

Let the proportion of marriages at any time where the woman is z to $z+dz$ years younger than the man be $\theta(z)\,dz$. $\theta(z)$ is in principle determinable from census data on the frequencies of marriages classified by the ages of the partners. Clearly, $\theta(z) = \phi(z)/\int_{-\infty}^{\infty} \phi(z)\,dz$. Let $C = A/B$ be the ratio of the rate of marriage to the rate of birth. Equations (2.54) thus yield

$$\phi(z) = \theta(z)C^2/a(1-a)A \qquad (2.55)$$

The calculation of Y_ξ. Equation (2.55) can be combined with equations (2.50), (2.52b) and (2.53) to obtain the expression

$$Y_\xi = C^2(V+\mu^2-\mu) \int_{-\infty}^{\infty} \theta(z)h_\xi(z)\,dz \qquad (2.56)$$

Y_ξ is thus independent of the absolute birth-rate and hence of the population size. The quantities V, μ, C and $\theta(z)$ can be determined from empirical data. It thus remains only to find an expression for $h_\xi(z)$ in terms of observable quantities. This can be achieved as follows. For concreteness, consider the relationship determined by the pedigree in Figure 2.2. The difference in age between the individuals of interest, J and I, is given by $z = (t+x_1+x_2)-(t+y+x_3) = x_1+x_2-(y+x_3)$. In order to calculate the probability density of z, $h_\xi(z)$, it is convenient to assume that the ages x_1, x_2 and x_3 of the ancestors 1, 2 and 3 of I and J are independently and normally distributed, as is the difference y

between the dates of birth of the sibs 1 and 3. Consider a set of infants born at the same instant. Let T and V_T be the mean and variance of the ages of their mothers, and T^* and V_T^* be the corresponding parameters for their fathers. Since the population is assumed to be stationary, we have (cf. p. 33)

$$T = \int_b^d xk(x)\,dx, \qquad V_T = \int_b^d (x - T)^2 k(x)\,dx \qquad (2.57a)$$

$$T^* = \int_{b*}^{d*} xk^*(x)\,dx, \qquad V_T^* = \int_{b*}^{d*} (x - T^*)^2 k^*(x)\,dx$$
$$(2.57b)$$

The means and variances of x_2 and x_3 are T and V_T respectively, while the mean and variance of x_1 are T^* and V_T^* respectively. The mean of y is equal to the mean difference between the ages of a randomly chosen pair of sibs, and is therefore zero. The variance of y, V_y, is equal to the variance of the difference in age between randomly chosen pairs of sibs. Hence, in this example the mean and variance of z are equal to

$$\mu_\xi = T + T^* - T = T^* \qquad (2.58a)$$

$$V_\xi = 2V_T + V_T^* + V_y \qquad (2.58b)$$

Figure 2.2. An imaginary pedigree. I and J are the individuals with relationship ξ. The individuals between the dotted lines are counted as ancestors of I and J. The expressions next to I, J and their ancestors represent their dates of birth; y is the difference in date of birth between the first ancestors of the male and female partners in the relationship ξ (J and I); x_1, x_2 and x_3 are the ages of individuals 1, 2 and 3 at the times of birth of their offspring displayed in the pedigree.

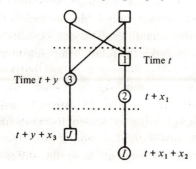

More generally, let there be α_ξ males and β_ξ females in the ancestry of the girl, and α_ξ^* males and β_ξ^* females in the ancestry of the boy involved in the relationship ξ. (The original ancestral couple are not counted as ancestors for this purpose.) The mean and variance of z are given by

$$\mu_\xi = (\alpha_\xi - \alpha_\xi^*)T^* + (\beta_\xi - \beta_\xi^*)T \qquad (2.59a)$$

$$V_\xi = (\alpha_\xi + \alpha_\xi^*)V_T^* + (\beta_\xi + \beta_\xi^*)V_T + V_y \qquad (2.59b)$$

The probability density function $h_\xi(z)$ is thus given by

$$h_\xi(z) = \frac{1}{\sqrt{2\pi V_\xi}} \exp\left[-\frac{(z-\mu_\xi)^2}{2V_\xi}\right] \qquad (2.60)$$

If we assume that the difference between the ages of spouses is also normally distributed with mean μ_a and variance V_a, we have

$$\theta(z) = \frac{1}{\sqrt{2\pi V_a}} \exp\left[-\frac{(z-\mu_a)^2}{2V_a}\right] \qquad (2.61)$$

Combining this with equation (2.60), substituting into equation (2.56) and evaluating the integral, we obtain the following expression for Y_ξ.

$$Y_\xi = \frac{C^2(V+\mu^2-\mu)}{\sqrt{2\pi S_\xi}} \exp\left[-\frac{\Delta_\xi^2}{2S_\xi}\right] \qquad (2.62)$$

where $\Delta_\xi = \mu_\xi - \mu_a$ and $S_\xi = V_\xi + V_a$.

Applications of the model. The application of equation (2.62) to specific populations requires estimates of the relevant means and variances. Human populations which are effectively isolated are generally examples of high-fertility, high-mortality populations, so that we need data on pre-industrial societies for these estimates. Hajnal (1963) suggests values of $T = 25$, $V_T = 40$, $T^* = 30$ and $V_T^* = 70$ for such societies. (Note that fathers have a considerably higher mean and variance than mothers.) Values of $\mu_a = 5$ and $V_a = 40$ are also suggested by Hajnal, who used data on an eighteenth-century French population to obtain $V + \mu^2 - \mu = 20$ and $V_y = 55$. C is the ratio of marriages to births, so that $2C$ is approximately equal to the proportion of people who survive to marriageable age. C may be taken as 0.25 for present purposes.

Armed with these figures, it is easy to calculate the expected rate of consanguineous marriages of any kind. Table 2.3 shows the values of the components of Y_ξ for the four classes of first-cousin marriage, from the above formulae, and Table 2.4 shows the

Table 2.3. *Some quantities required in the calculation of Y_ξ for the four kinds of first-cousin marriage*

Class ξ	$\alpha_\xi + \alpha_\xi^*$	$\beta_\xi + \beta_\xi^*$	S_ξ	Δ_ξ
Paternal parallel cousin	2	0	225	$-\mu_a$
Maternal parallel cousin	0	2	165	$-\mu_a$
Patrilateral cross cousin	1	1	195	$-2\mu_a$
Matrilateral cross cousin	1	1	195	0

Table 2.4. *Theoretical frequencies of the four kinds of first-cousin marriage (relative to a frequency of 100 for all first-cousin marriages)*

	Parallel		Cross	
μ_a	Paternal	Maternal	Patrilateral	Matrilateral
0	23.2	27.1	24.9	24.9
2	23.3	27.1	24.3	25.3
5	24.1	27.5	21.1	27.3
10	25.6	27.6	12.3	34.4

After Table 27 of Hajnal (1963)

relative frequencies of these marriages as a function of the mean difference in age between spouses. With $\mu_a = 5$, the sum of the Y_ξ is equal to 0.14; smaller values of μ_a give a slightly bigger, and larger ones a slightly smaller, value. Thus in an isolated population with high fertility, there would be between 1 and 1.5 first-cousin marriages per decade. This agrees well with data on the frequencies of first-cousin marriages in some nineteenth-century European isolates (Hajnal, 1963). Uncle–niece and aunt–nephew marriages taken together would be expected to occur at about 0.06 of this rate, marriages between first-cousins once removed at 0.6, and second cousins at 3.0, times this rate.

The reason for the low frequency of paternal parallel first-cousin marriages is clear from inspection of Table 2.3. Although the term arising from the mean difference between the ages of spouses is the same for paternal parallel and maternal parallel marriages, both ancestors in the paternal parallel pedigree are males, whereas both are female in the maternal parallel pedigree. The variance term S_ξ is therefore bigger in the former case; since Y_ξ decreases with S_ξ, there is a corresponding deficit in the frequency of marriages between paternal parallel first cousins. This result may be viewed intuitively as follows. There is greater variation in the age of fathers than that of mothers, so that paternal parallel cousins are more likely to be born far apart in time, and hence are less likely to marry each other. Conversely, maternal parallel cousins are the most likely of the four classes to be born close in time; this explains why maternal parallel cousin marriages are the most common, except when μ_a is so big that its effect on Δ_ξ overcomes the differences due to S_ξ (cf. Tables 2.3 and 2.4).

Comparison of Tables 2.2 and 2.4 shows that the theory seems to predict the qualitative pattern of the differences in frequencies of classes of first cousin marriages. Furthermore, Japan and Italy are known to have larger values of μ_a than Germany and Austria, so that the differences between the top and bottom of Table 2.2 accord well with the differences between the rows of Table 2.4. The differences between the theoretical frequencies do not seem as big as the actual ones. Hajnal (1963) discusses some possible reasons for these discrepancies.

Conclusions. The results derived here show that both the absolute rates and relative frequencies of consanguineous marriages in isolated populations are highly dependent on several demographic parameters. Changes in demography, therefore, may themselves produce shifts in marriage patterns. For example, a reduction in mortality would result in an increase in C, the ratio of the marriage rate to the birth rate, leading to an increase in both the absolute rate and frequency of consanguineous marriages (since Y_ξ is proportional to C^2). The model predicts a constant annual rate of consanguineous marriages in a stationary population, independent of the population size, which seems to agree well with observations on human populations. Hajnal (1963) has also investigated the effects of population growth and migration, and Cavalli-Sforza *et al.* (1966) have carried out an elaborate investigation of the consequences of migration. Although some quantitative differences are introduced by these factors, the general pattern of results is not affected very much. The relative frequencies of different classes of consanguineous marriages are expected to be similar to those for a stationary population, unless the rate of population growth is very high, but the frequency of consanguineous marriages as a whole will decrease as the population size increases. The effect of migration is to lower the rate of consanguineous marriages; if male and female migration rates differ, the pattern of consanguineous marriages may be affected. Since males often have higher migration rates than females in man, one might expect migration to lower the frequency of marriages that involve a relatively large number of male ancestors (e.g. between paternal parallel cousins), and to raise the frequency of marriages involving a relatively large number of female ancestors (e.g. between maternal parallel cousins). This may help to explain the disagreements between Tables 2.2 and 2.4.

Overall, data on the frequencies of consanguineous marriages in human populations suggest that the assumption of random mating is close to the truth (Hajnal, 1963), so that people do not appear to select their spouses on the basis of relationship (except for the avoidance of incest). It is interesting to note that phenomena such as the relative infrequency of paternal parallel cousin marriages have sometimes been interpreted as evidence for such selection. Analyses of this sort can, of course, in principle be carried out for non-human populations, although data of a suitable sort are rarely available. Hajnal (1976) has analysed data on an English population of great tits, and shown that the observed frequencies of consanguineous marriages are consistent with random mating.

Finally, it is useful to note how these results can be used to estimate the size of isolates, if we have data on populations where it can be assumed that no migration occurs between random-mating units. If we know that there are A_ξ marriages of class ξ per year in the region of interest, but we expect Y_ξ in an isolated population, the number of isolates is given by A_ξ/Y_ξ. If f_ξ is the frequency of marriages of class ξ, then the number of marriages per isolate per year is equal to $A/(A_\xi/Y_\xi) = Y_\xi/f_\xi$. If $\tilde{A}$ is the average number of marriages per year per head of population (counting individuals of all ages), and N is the size of an isolate, we have $\tilde{A}N = Y_\xi/f_\xi$, so that

$$N = Y_\xi/\tilde{A}f_\xi \tag{2.63}$$

This replaces Dahlberg's (1929, 1948) formulae.

3

Selection: construction of a model and the properties of equilibrium populations

3.1 Introduction

The problem of the theory of natural selection in age-structured populations was first considered quite early in the history of population genetics, by Haldane (1927a) and Norton (1928). Until comparatively recently, there has been little interest in adding to the important contributions of these authors, apart from the isolated paper of Haldane (1962). This is probably largely due to the introduction by Fisher (1930) of his well-known 'Malthusian parameter' method of dealing with selection in continuous-time populations, which apparently provided a simple and elegant answer to the problem, and has since been extensively employed in population genetics (Crow and Kimura, 1970). This method assumes that fixed *per capita* birth-rates and death-rates can be assigned to each genotype, analogous to the rates defined in section 1.3.2 for a genetically homogeneous population in stable age-distribution. The Malthusian parameter for a genotype is the difference between these two quantities, and is used in differential equations for gene frequency change in much the same way that the standard discrete-generation fitness measure of Wright and Haldane is used in difference equations. But as pointed out by several authors (Moran, 1962; Charlesworth, 1970; Pollak and Kempthorne, 1970, 1971), an individual genotype cannot have a fixed *per capita* birth-rate and death-rate when the genotypic composition of a population is changing under natural selection, resulting in a continual disturbance of the population's demographic structure.

The question of the analysis of the process of selection in age-structured populations is, therefore, not satisfactorily settled by the Malthusian parameter technique, and interest in the problem has revived in recent years, as can be seen from the publications of

118

Anderson and King (1970), Charlesworth (1970, 1972–4, 1976), Pollak and Kempthorne (1970, 1971), King and Anderson (1971), Charlesworth and Giesel (1972*a*, *b*) and Charlesworth and Charlesworth (1973). This has coincided with strong interest among evolutionary theorists in the problem of the evolution of life-histories (i.e. the phenotypic consequences, at the level of the $l(x)$ and $m(x)$ functions, of selection in age-structured populations); this topic is discussed in the final chapter.

The present chapter is concerned with the construction of a reasonably realistic and rigorous model of the population genetics of natural selection, and with the analysis of genetic equilibria under selection. The following chapter deals with the dynamic aspects of selection. Both finish with a discussion of the biological implications of the main results.

The mathematical models throughout these chapters are developed mainly in terms of the difference equation approach to age-structured populations, described in section 1.2.1, since in studying the selection process this method gives a more compact notation for the highly non-linear equations which arise than the equivalent matrix formulation. It also permits an easy extension to the integral equations of the corresponding continuous-time model. Only sexually reproducing populations are considered; the relatively trivial case of selection in an asexual population has already been dealt with in section 1.5. The presentation is for the case of two distinct sexes, although hermaphrodite and monoecious species can be described by the same equations if male and female gametes from the same individual are considered separately. Random mating with respect to genotype is assumed throughout (complete self-fertilisation has been studied by Pollak and Kempthorne, 1970). Selection is assumed throughout to be frequency independent, although considerable attention is paid to density-dependent selection (cf. Anderson, 1971; Charlesworth, 1971; Roughgarden, 1971, 1976; Kimura, 1978; Nagylaki, 1979*a*). This is not because I believe that frequency-dependent selection is unimportant, but because the aim of this study is to see to what extent age-structure introduces novel features into the process of selection; the introduction of frequency dependence would complicate further an already complex problem. A theoretical study of one type of frequency-dependent selection, Batesian mimicry, has been made in age-structured context by Charlesworth and Charlesworth (1975).

3.2   Construction of a model of selection

3.2.1   *Genotypic parameters*

As in the case of selection in an asexual species (section 1.5.2), it is necessary to specify demographic parameters for each genotype. For concreteness, consider an autosomal locus with alleles $A_1, A_2, \ldots A_n$, and let A_iA_j be the genotype of an individual who received A_i from its mother and A_j from its father. Reciprocal effects are assumed to be absent, so that the parameters of A_iA_j and A_jA_i are identical. Let $P_{ij}(x, t)$ and $P_{ij}^*(x, t)$ be the probabilities of survival from age-class x at time t to age-class $x + 1$ at time $t + 1$ for A_iA_j females and males respectively. As in section 1.2.1, we can derive from these the probabilities of A_iA_j females and males surviving to age x at time t from conception at time $t - x$

$$l_{ij}(x, t) = \prod_{y=1}^{x} P_{ij}(x - y, t - y) \tag{3.1a}$$

$$l_{ij}^*(x, t) = \prod_{y=1}^{x} P_{ij}^*(x - y, t - y) \tag{3.1b}$$

In principle, we can also define the fecundity of an A_iA_j individual aged x at time t in terms of the number of offspring it is expected to produce in time-interval t. By analogy with section 1.2.1, this can be written as $M_{ij}(x, t)$ for a female and $M_{ij}^*(x, t)$ for a male. Empirically, these fecundities can be measured simply by estimating the mean number of offspring attributable to an A_iA_j female or male of age x at time t. As in the case of a genetically homogeneous population, these parameters cannot in general be treated as independent of the age-composition of the population. The problem is even worse in the present case since, if different classes of matings have different fecundities, the net fecundity of individuals of a given genotype, age and sex may be affected by the genotypic composition of the population of the opposite sex, as well as by its age composition. This problem also arises, in a less acute form, in discrete generation models of selection (Bodmer, 1965; Kempthorne and Pollak, 1970; Prout, 1971a, b). It is dealt with in that case by making specialising assumptions about the way in which selection on fecundity works, and this is essentially what will be done here.

3.2.2 *Models of the mating process*

It is convenient and realistic to make the assumption used in deriving equation (1.5): the expected number of offspring produced at time t by an A_iA_j female aged x is independent of the genotypes and ages of her mate or mates. Hence, $M_{ij}(x, t)$ is independent of the composition of the population with respect to both age and genotype, unless frequency-dependent selection is operating.

If $N_{ij}(x, t)$ is the number of A_iA_j females aged x at time t, the total number of zygotes produced at time t is

$$B(t) = \sum_{ijx} N_{ij}(x, t)M_{ij}(x, t) \tag{3.2a}$$

Similarly, let $N_{ij}^*(x, t)$ be the number of A_iA_j males aged x at time t. Since every individual has a mother and father, we must also have

$$B(t) = \sum_{ijx} N_{ij}^*(x, t)M_{ij}^*(x, t) \tag{3.2b}$$

The summations in these equations are taken over all reproductively active individuals. Selection may well involve the ages of first and last reproduction; in order to avoid excessive use of subscripts, wherever upper and lower ages of reproduction appear (b and d for females, b^* and d^* for males) these are to be taken as representing the maximum upper and minimum lower values respectively for all genotypes, unless genotypes are specifically identified by subscripts.

Since $M_{ij}(x, t)$ is assumed to be independent of the population's composition, equations (3.2) imply that the absolute values of the $M_{ij}^*(x, t)$ are constrained by the state of the male and female populations. The nature of this constraint is particularly simple if a second assumption is introduced: that mating of fertile individuals is random with respect to both genotype and age. Such randomness arises naturally in the following two models of the mating process.

Sperm or pollen pool model. In this model, breeding males at time t are assumed to shed their gametes into a common pool of effectively infinite size, from which random draws are made to fertilise the ova. Let $\theta_{ij}(x, t)$ be the expected number of gametes contributed to the pool at time t by an $A_i A_j$ male aged x. This quantity characterises the absolute fecundity of the class of male in question in the same way that the $M_{ij}(x, t)$ functions characterise female fecundity. The $\theta_{ij}(x, t)$ may in principle be independent of the state of the population. The total size of the male gamete pool at time t is

$$B^*(t) = \sum_{ijx} N^*_{ij}(x, t)\theta_{ij}(x, t) \qquad (3.3)$$

Only $B(t)$ of these gametes actually fertilise ova, so that the net fecundity of an $A_i A_j$ male aged x is

$$M^*_{ij}(t, x) = \theta_{ij}(x, t)B(t)/B^*(t) \qquad (3.4)$$

where $B(t)$ is determined by equation (3.2a).

This model of mating is realistic for organisms with external fertilisation, such as many species of marine animals which shed their sperm and eggs into the sea, and self-incompatible higher plants, whose pollen is dispersed by wind or insect vectors. For organisms with internal fertilisation, the following model is more appropriate.

Mating group model. During time-interval t, females mate with males chosen by draws from a group composed of all fertile males. Independent draws are made from the group until all the fertile females present at time t have successfully mated. The quantity $\theta_{ij}(x, t)$ which characterises a male of a given class is now to be regarded as the probability that he is chosen in a draw from the group. The *relative* values of the $\theta_{ij}(x, t)$ are independent of the composition of the population unless there is frequency-dependent selection. The net fecundities $M^*_{ij}(x, t)$ can be calculated by a technique similar to that used for the gamete pool model, yielding an expression identical with equation (3.4). $B^*(t)$ in this case is defined purely formally, however, and has no concrete biological meaning.

This model provides a reasonably accurate description of, for example, many species of insect or small vertebrate. It does not apply to species such as man, where there is a high correlation between the ages of mates. An examination of the consequences of non-random mating with respect to age is made later in this chapter (section 3.2.4), but for the present the assumption of random mating will be retained.

3.2.3 Genotypic frequencies

With the single-locus case and random mating, it is easily seen that the genotypic frequencies among the zygotes produced at time t are identical with those obtained by combining pairs of alleles drawn at random from a pool of maternally derived genes and a pool of paternally derived genes. As in section 2.2.1, let the frequencies of allele A_i among the maternal and paternal pools be $p_i(t)$ and $p_i^*(t)$ respectively. The frequency of A_iA_j among the zygotes is

$$p_{ij}(t) = p_i(t)p_j^*(t) \tag{3.5}$$

where the allele frequencies are given by

$$B(t)p_i(t) = \tfrac{1}{2}\sum_{jx}[N_{ij}(x, t) + N_{ji}(x, t)]M_{ij}(x, t) \tag{3.6a}$$

$$B(t)p_i^*(t) = \tfrac{1}{2}\sum_{jx}[N_{ij}^*(x, t) + N_{ji}^*(x, t)]M_{ij}^*(x, t) \tag{3.6b}$$

These equations can be simplified by assuming that the frequency of females among new zygotes has a fixed value a, which is independent of the ages and genotypes of the parents. The number of male and female zygotes can then be written as fixed fractions of the total number of zygotes, $aB(t)$ and $(1-a)B(t)$ respectively. We can also write $m_{ij}(x, t)$ for the expected number of daughters of an A_iA_j female, and $m_{ij}^*(x, t)$ for the expected number of sons of an A_iA_j male, aged x at time t.

$$m_{ij}(x, t) = aM_{ij}(x, t) \tag{3.7a}$$

$$m_{ij}^*(x, t) = (1-a)M_{ij}^*(x, t) \tag{3.7b}$$

This enables us to define reproductive functions for each genotype as

$$k_{ij}(x, t) = l_{ij}(x, t)m_{ij}(x, t) \tag{3.8a}$$

$$k_{ij}^*(x, t) = l_{ij}^*(x, t)m_{ij}^*(x, t) \tag{3.8b}$$

(cf. section 1.2.1). For $t \geqq x$, we have from equation (3.5)

$$N_{ij}(x, t) = aB(t-x)p_i(t-x)p_j^*(t-x)l_{ij}(x, t) \tag{3.9a}$$

$$N_{ij}^*(x, t) = (1-a)B(t-x)p_i(t-x)p_j^*(t-x)l_{ij}^*(x, t) \tag{3.9b}$$

These relationships can be used to formulate difference equations for the allele frequencies. When $t \geqq d, d^*$, all the zygotes in the population are produced by parents who were born after the initial time-interval. Equations (3.6) become, using equations (3.8) and (3.9),

$$B(t)p_i(t) = \tfrac{1}{2} \sum_{jx} B(t-x)[p_i(t-x)p_j^*(t-x) \\ + p_j(t-x)p_i^*(t-x)]k_{ij}(x, t) \tag{3.10a}$$

$$B(t)p_i^*(t) = \tfrac{1}{2} \sum_{jx} B(t-x)[p_i(t-x)p_j^*(t-x) \\ + p_j(t-x)p_i^*(t-x)]k_{ij}^*(x, t) \tag{3.10b}$$

where

$$B(t) = \sum_{ijx} B(t-x)p_i(t-x)p_j^*(t-x)k_{ij}(x, t)$$

$$= \sum_{ijx} B(t-x)p_i(t-x)p_j^*(t-x)k_{ij}^*(x, t) \tag{3.10c}$$

For times earlier than this, contributions from individuals alive in the initial time-interval must be taken into account as well. We use a method similar to that of equation (1.18). Write

$$g_i(t) = \tfrac{1}{2} \sum_{x=t+1}^{d} \sum_{j} [N_{ij}(x-t, 0) + N_{ji}(x-t, 0)]\tilde{l}_{ij}(x, t)M_{ij}(x, t) \tag{3.11a}$$

$$g_i^*(t) = \tfrac{1}{2} \sum_{x=t+1}^{d} \sum_{j} [N_{ij}^*(x-t, 0) + N_{ji}^*(x-t, 0)]\tilde{l}_{ij}^*(x, t)M_{ij}^*(x, t) \tag{3.11b}$$

where $g_i(t) = 0$ for $t \geq d$, $g_i^*(t) = 0$ for $t \geq d^*$; $\tilde{l}_{ij}(x, t)$ and $\tilde{l}_{ij}^*(x, t)$ are defined analogously to $\tilde{l}(x, t)$ in equation (1.18). It is easily seen that the full versions of equations (3.10) for arbitrary t, are

$$B(t)p_i(t) = g_i(t) + \tfrac{1}{2} \sum_{x=1}^{t} B(t-x) \sum_j [p_i(t-x)p_j^*(t-x)$$

$$+ p_j(t-x)p_i^*(t-x)]k_{ij}(x, t) \tag{3.12a}$$

$$B(t)p_i^*(t) = g_i^*(t) + \tfrac{1}{2} \sum_{x=1}^{t} B(t-x) \sum_j [p_i(t-x)p_j^*(t-x)$$

$$+ p_j(t-x)p_i^*(t-x)]k_{ij}^*(x, t) \tag{3.12b}$$

$$B(t) = \sum_i g_i(t) + \sum_{x=1}^{t} B(t-x) \sum_{ij} p_i(t-x)p_j^*(t-x)k_{ij}(x, t) \tag{3.12c}$$

$$B(t) = \sum_i g_i^*(t) + \sum_{x=1}^{t} B(t-x) \sum_{ij} p_i(t-x)p_j^*(t-x)k_{ij}^*(x, t) \tag{3.12d}$$

There are $2(n+1)$ equations in this system, taking i over all possible values, but only $2n$ are independent, since $\sum p_i = \sum p_i^* = 1$. These equations can be simplified further if it is assumed that the life-tables for each genotype are identical for males and females, and that the age-specific fecundity functions for each genotype have the same shape with respect to age in males and females, i.e. if we have

$$l_{ij}(x, t) = l_{ij}^*(x, t) \tag{3.13a}$$

$$\theta_{ij}(x, t) \propto M_{ij}(x, t) \tag{3.13b}$$

(Note that this implies $b = b^*$ and $d = d^*$.)

If these relations are assumed, it is easy to see that $p_i(t) = p_i^*(t)$ for $t \geq d$, regardless of the initial conditions. If we shift the time-scale so that the initial time interval falls in the period when $p_i = p_i^*$, a full description of the population's state is given by the reduced system of $n+1$ equations, n of which are independent

$$B(t)p_i(t) = g_i(t) + \sum_{x=1}^{t} B(t-x) \sum_j p_i(t-x)p_j(t-x)k_{ij}(x, t) \tag{3.14a}$$

$$B(t) = \sum_i g_i(t) + \sum_{x=1}^{t} B(t-x) \sum_{ij} p_i(t-x)p_j(t-x)k_{ij}(x, t) \tag{3.14b}$$

The assumptions embodied in equations (3.13) are somewhat arbitrary, and are unlikely to hold for many biological situations. Equations (3.14) are useful, however, since they resemble closely the standard discrete-generation selection equations of population genetics (Crow and Kimura, 1970, p. 180), which in fact may be regarded as the special case of equations (3.14) with $b = d = 1$; they take the form

$$\bar{W}(t)p_i(t) = p_i(t-1) \sum_j p_i(t-1) W_{ij}(t-1) \tag{3.15a}$$

$$\bar{W}(t) = \sum_{ij} p_i(t-1)p_j(t-1) W_{ij}(t-1) \tag{3.15b}$$

where $W_{ij}(t)$ is the fitness of A_iA_j in generation t.

Equations (3.14) can therefore be used to compare the properties of age-structured populations with those of discrete-generation populations, as well as being more tractable than equations (3.12) for mathematical analysis. Equations of this form, or their matrix or continuous-time equivalents (equations (3.16)) have commonly been used in studies of the theory of selection in age-structured populations (e.g. Haldane, 1927a, 1962; Norton, 1928; Anderson and King, 1970; Charlesworth, 1970, 1972, 1976; Pollak and Kempthorne, 1971). The numerous assumptions needed to arrive at these equations are summarised in Table 3.1, and should be

Table 3.1. *Assumptions involved in deriving the difference equations for gene frequencies*

Equations (3.12) and (3.14) require that:
  (1) The fecundity of a mating between a given pair of individuals is determined solely by the age and genotype of the female.
  (2) Mating is at random with respect to age and genotype.
  (3) The primary sex-ratio is constant and independent of parental age and genotype.

Equations (3.14) also require that:
  (4) The age-specific survival probabilities of males and females of a given genotype are equal.
  (5) The age-specific fecundity functions of males and females of a given genotype are proportional.

---

borne in mind whenever they are used. The earlier papers on this subject were rather casual about the nature of their assumptions concerning the mating process and mode of selection; Pollak and Kempthorne (1971) were the first to emphasise the need for clarity on these points. They in fact went further than we have done and assumed that the fecundities of each genotype were the same and were independent of age for each fertile age-class. These assumptions are not necessary, and their general results are valid for the model leading to equations (3.14). For further details of the models described in this section, see Charlesworth (1972, 1976).

The continuous-time model. These results can easily be extended to the continuous-time case by increasing indefinitely the number of age-classes within the life-span. As in section 1.2.2, the population size at time t is characterised by the population birth-rate $B(t)$, such that the number of zygotes produced in a short time-interval of length dt is $B(t)\,dt$. The frequency of A_i among maternally-derived genes of new zygotes at time t is defined as $p_i(t)$, such that the number of A_i genes of maternal origin among the zygotes produced between t and $t+dt$ is $B(t)p_i(t)\,dt$; the paternal allele frequency, $p_i^*(t)$, is defined similarly. By re-defining $l_{ij}(x, t)$, $m_{ij}(x, t)$ and $k_{ij}(x, t)$, etc., appropriately and re-formulating the mating models in terms of arbitrarily small time-intervals, integral equations analogous to (3.12) and (3.14) can be obtained. For example, the integral analogues of equations (3.14) are

$$B(t)p_i(t) = g_i(t) + \int_0^t B(t-x)p_i(t-x) \sum_j p_j(t-x)k_{ij}(x, t)\,dx \quad (3.16a)$$

$$B(t) = \sum_i g_i(t) + \int_0^t B(t-x) \sum_{ij} p_i(t-x)p_j(t-x)k_{ij}(x, t)\,dx \quad (3.16b)$$

$$g_i(t) = \tfrac{1}{2} \int_t^d \sum_j [N_{ij}(x-t, 0) + N_{ji}(x-t, 0)]\tilde{l}_{ij}(x, t)M_{ij}(x, t)\,dx \quad (3.16c)$$

where $\tilde{l}_{ij}(x, t)$ is defined analogously to $\tilde{l}(x, t)$ in equation (1.26).

Cornette (1975) and Nagylaki (1977, pp. 79–82) should be consulted for an alternative formulation of the mating process with the continuous-time model.

The two-locus case. The discrete age-class model developed above can easily be extended to genetic systems more complex than that of a single autosomal locus. For example, consider a two-locus system with alleles A_1, A_2 at one locus and B_1, B_2 at the other. With the assumptions that led to equations (3.14), it is easy to derive difference equations which describe the system completely. Let c be the recombination fraction, and write the frequencies of the gametes A_1B_1, A_1B_2, A_2B_1 and A_2B_2 at time t as $p_1(t)$, $p_2(t)$, $p_3(t)$ and $p_4(t)$, respectively. The coefficient of linkage disequilibrium (cf. p. 86) is $D(t) = p_1(t)p_4(t) - p_2(t)p_3(t)$. For $t \geqq d$ we have

$$B(t)p_i(t) = \sum_{jx} B(t-x)p_i(t-x)p_j(t-x)k_{ij}(x, t)$$

$$\pm \sum_x cB(t-x)D(t-x)k_{14}(x, t) \qquad (3.17a)$$

$$B(t) = \sum_{ijx} B(t-x)p_i(t-x)p_j(t-x)k_{ij}(x, t) \qquad (3.17b)$$

where the sign attached to c is positive for $i = 2, 3$ and negative for $i = 1, 4$.

The extension of these equations to times $t < d$ is trivial.

3.2.4 *Non-random mating with respect to age*

Before going on to explore the properties of the gene frequency equations derived in the previous sections of this chapter, it is worthwhile to consider the consequences of relaxing the assumption of randomness of mating with respect to age. The present account is based on that of Charlesworth and Charlesworth (1973). It is necessary in this case to consider all possible matings with respect to the ages of the partners. Let $\psi(x, y, t)$ be the frequency of matings at time t involving females aged x and males aged y. Let $p_i(x, t)$ be the frequency of A_i among the gametes produced by females in such matings, and $p_i^*(y, t)$ be the corresponding frequency among male gametes. It is convenient to write

$$\bar{p}_i(x, y, t) = \tfrac{1}{2}[p_i(x, t) + p_i^*(y, t)] \qquad (3.18a)$$

$$\delta_i(x, y, t) = p_i(x, t) - \bar{p}_i(x, y, t) = \bar{p}_i(x, y, t) - p_i^*(y, t) \qquad (3.18b)$$

If there is random mating with respect to genotype, the genotypic frequencies among the zygotes produced by this class of matings may be written (dropping the argument t for brevity)

$$p_{ii}(x, y) = p_i(x)p_i^*(y) = \bar{p}_i^2(x, y) - \delta_i^2(x, y) \qquad (3.19a)$$

$$p_{ij}(x, y) + p_{ji}(x, y) = p_i(x)p_j^*(y) + p_j(x)p_i^*(y)$$
$$= 2\bar{p}_i(x, y)\bar{p}_j(x, y) - 2\delta_i(x, y)\delta_j(x, y) \qquad (3.19b)$$

Taking expectations over all classes of mating, we find that the overall genotypic frequencies among the new zygotes are given by

$$p_{ii} = \sum_{xy} \psi(x, y)[\bar{p}_i^2(x, y) - \delta_i^2(x, y)]$$
$$= \bar{p}_i^2 - \bar{\delta}_i^2 + V[\bar{p}_i(x, y)] - V[\delta_i(x, y)] \qquad (3.20a)$$

$$p_{ij} + p_{ji} = 2 \sum_{xy} \psi(x, y)[\bar{p}_i(x, y)\bar{p}_j(x, y) - \delta_i(x, y)\delta_j(x, y)]$$
$$= 2\bar{p}_i\bar{p}_j - 2\bar{\delta}_i\bar{\delta}_j + 2\ Cov\ [\bar{p}_i(x, y), \bar{p}_j(x, y)]$$
$$- 2\ Cov\ [\delta_i(x, y), \delta_j(x, y)] \qquad (3.20b)$$

where $\bar{p}_i = \sum_{xy} \psi(x, y)\bar{p}_i(x, y)$, $\bar{\delta}_i = \sum_{xy} \psi(x, y)\delta_i(x, y)$, and the V and Cov functions are the variances and covariances across age-classes of the variables enclosed in square brackets.

If selection is not intense, we can ignore the second-order terms in the differences in the gene frequencies and in the δ_i between different classes of mating (section 4.2), so that the variance and covariance terms can be dropped from equations (3.20), to a good approximation. Noting that $p_i(t) = \bar{p}_i(t) + \bar{\delta}_i(t)$ and $p_i^*(t) = \bar{p}_i(t) - \delta_i(t)$, the approximate expressions can be substituted into equations (3.9), enabling us to recover equations (3.10) and (3.12). This shows that, as far as the formal expressions for the gene frequencies are concerned, non-random mating with respect to age has only a second-order effect. But, unlike the random-mating case (section 3.2.2), it is no longer possible to define $\theta_{ij}(x, t)$ functions which are independent of the composition of the population. This means that with non-random mating between age-classes, equations (3.10) etc. are useful only in treating equilibrium populations, and certain restricted types of non-equilibrium situations to be discussed in section 4.3.1.

With intense selection, equations (3.20) show that deviations from Hardy–Weinberg frequencies among the new zygotes can be produced even if there is random mating with respect to genotype. With just two alleles, A_1 and A_2, equations (3.20) become

$$p_{11} = \bar{p}_1^2 - \bar{\delta}_1^2 + V[\bar{p}_1(x, y)] - V[\delta_1(x, y)] \qquad (3.21a)$$

$$p_{12} + p_{21} = 2\bar{p}_1\bar{p}_2 + 2\bar{\delta}_1^2 - 2V[\bar{p}_1(x, y)] + 2V[\delta_1(x, y)] \qquad (3.21b)$$

The effects of differences in gene frequency between the sexes and differences between classes of mating are in opposition, the former tending to produce an excess of heterozygotes, and the latter a deficiency, over the Hardy–Weinberg frequencies. If selection acts in a similar way in males and females, it is probable that the terms in δ_1^2 will be small, compared with the variance in gene frequencies over classes of matings, resulting in a net deficiency of heterozygotes among the new zygotes. Of course, if there is a viability advantage of the heterozygote, this deficiency of heterozygotes will diminish among older age-classes and may be replaced by an excess of heterozygotes. This may be relevant to some observations of heterozygote deficiencies in natural populations (e.g. Schaal and Levin, 1976).

3.3   Populations in genetic equilibrium

In this section we examine the more important properties of age-structured populations in genetic equilibrium under selection. For convenience, the simplified selection equations (3.14) are used for deriving most of the results, although reference to the more realistic models will be made where appropriate. The treatment here is derived largely from the papers of Charlesworth (1972) and Charlesworth and Charlesworth (1973).

3.3.1   *Dependence of genetic equilibrium on demographic stability*

One of the most important results in the theory of selection in age-structured populations is that genetic equilibrium is generally possible only if there is demographic stability, i.e. if the population is stationary in size or growing at a constant rate, and has a constant age-structure. A corollary of this is that populations which are subject to fluctuations in growth-rate and age-structure may show fluctuations in their composition at loci undergoing selection (Anderson and King, 1970; Charlesworth and Giesel, 1972a, b). The biological implications of this result are discussed in section 3.4.1; here we will investigate its mathematical basis.

In an equilibrium population, the gene frequencies at a locus must take values independent of time, so we can write $p_i(t) = \hat{p}_i$ in equations (3.14), yielding the equilibrium equations for $t \geqq d$

$$1 = \sum_j \hat{p}_j \sum_x \frac{B(t-x)}{B(t)} k_{ij}(x, t) \tag{3.22a}$$

$$B(t) = \sum_{ijx} B(t-x)\hat{p}_i\hat{p}_j k_{ij}(x, t) \tag{3.22b}$$

These may be compared with the equilibrium form of the discrete-generation equations (3.15), writing the fitnesses W_{ij} as independent of time, as must usually be the case in an equilibrium population

$$1 = \sum_j \hat{p}_j \frac{W_{ij}}{\hat{W}} \tag{3.23a}$$

$$\hat{W} = \sum_{ij} \hat{p}_i\hat{p}_j W_{ij} \tag{3.23b}$$

Only the relative values of the W_{ij} in equations (3.23) are important in determining the equilibrium gene frequencies. If we wish to define a fitness measure for the age-structured case which determines the equilibrium gene frequencies in the same way as the W_{ij} in the discrete-generation case, we can therefore choose a scale in which the equivalent of $\hat{W}$ is 1, with no loss of generality; this yields the following equilibrium fitness measure for the age-structured case

$$w_{ij} = \sum_x \frac{B(t-x)}{B(t)} k_{ij}(x, t) \tag{3.24}$$

(The fitness measure is given in lower case to distinguish it from the discrete-generation W_{ij}.)

For there to be genetic equilibrium, the relative values of these w_{ij} must be independent of time, except in certain degenerate cases discussed by Charlesworth (1972). Since variation with t of the function $B(t-x)/B(t)$ affects the relative contributions of different ages to the w_{ij}, it is difficult to imagine conditions under which a population with a varying growth-rate, and hence varying $B(t-x)/B(t)$, could be in genetic equilibrium, unless we can write for all ij

$$k_{ij}(x, t) = w'_{ij}k_s(x, t) \qquad (3.25)$$

where $k_s(x, t)$ is the reproductive function of some standard geno-type, and w'_{ij} is a constant independent of x and t but characteristic of A_iA_j.

It is unlikely that equation (3.25) can often be satisfied. If selection acts at the level of age-specific survival probabilities, for example, and if genotypic differences in these are independent of age, we can write

$$l_{ij}(x) = P^x_{ij}l_s(x)$$

where $l_s(x)$ is the survival function for the genotype with the highest survival probabilities, and $P_{ij} \leqq 1$ is a constant characteristic of A_iA_j. Even in this simple case, equation (3.25) is violated. Similarly, if selection acts at the level of fecundity, it is likely that genotypes will differ in such parameters as age at first or last reproduction, time of peak fecundity etc., which will also be inconsistent with equation (3.25).

We may conclude, therefore, that a population will be in genetic equilibrium only if the population has a constant growth-rate and hence stable age-structure, unless selective differences among genotypes are non-specific with respect to age in such a way that equation (3.25) is satisfied. Of course, deviations from genetic equilibrium caused by changing age-structure may be so slight as not to be noticeable in practice; the conditions under which there are significant effects of changing demography are discussed in section 3.4.1.

Proof for the time-independent case. A formal proof that demo-graphic stability is necessary for genetic equilibrium under selection can be given in the case when the $k_{ij}(x, t)$ functions are independent of t. We assume that the population is at genetic equilibrium but does not possess a stable age-structure, and show that this leads to a contradiction except when equation (3.25) is satisfied.

If the population is in genetic equilibrium, it can be treated from the demographic viewpoint as if it were genetically homogeneous, replacing the usual reproductive function $k(x)$ (equation (1.16)) by the mean for the equilibrium population

$$k(\hat{p}, x) = \sum_{ij} \hat{p}_i \hat{p}_j k_{ij}(x) \tag{3.26}$$

From the standard theory described in section 1.3.1, it follows that, for a population with changing age-structure, we can write $B(t)$ at any time as

$$B(t) = \sum_{l=1}^{d} C_l \lambda_l^{t+1} \tag{3.27}$$

where the λ_l are the roots of the dth degree polynomial

$$\sum_x \lambda^{-x} k(\hat{p}, x) = 1 \tag{3.28}$$

(Equation (3.27) assumes for simplicity that the λ_l are all distinct, but the following proof can easily be extended to cover the cases of multiple roots.)

If the expression for $B(t)$ in equation (3.27) is substituted into equation (3.24), the following condition for constant w_{ij} must hold for each genotype and each value of t

$$\sum_{l=1}^{d} C_l \lambda_l^{t+1} \sum_x \lambda_l^{-x} k_{ij}(x) = w_{ij} \sum_{l=1}^{d} C_l \lambda_l^{t+1} \tag{3.29}$$

By choosing an arbitrary set of values of t, $[t_1, t_2, \ldots t_d]$, and substituting them into equation (3.29), we obtain a set of homogeneous linear equations in d unknowns, $[y_1, y_2, \ldots y_d]$

$$\sum_{l=1}^{d} a_{ml} y_l = 0 \tag{3.30}$$

where

$$y_l = \sum_x \lambda_l^{-x} \frac{k_{ij}(x)}{w_{ij}} - 1, \qquad a_{ml} = C_l \lambda_l^{t_m+1}$$

But, since the choice of t values is arbitrary, it must be possible for the determinant of the a_{ml} to be non-zero, which implies that the y_l must all be zero for equation (3.30) to hold. Each λ_l must therefore be a root of the following equation, for every i and j

$$\sum \lambda_l^{-x} \frac{k_{ij}(x)}{w_{ij}} = 1$$

Since the coefficients of a polynomial equation are completely determined by their roots, this implies that the $k_{ij}(x)/w_{ij}$ are the same for each ij and a given value of x, so that equation (3.25) is satisfied. Under any other conditions, the population cannot simultaneously be in genetic equilibrium and not have a stable age-structure.

3.3.2   *Equilibrium fitness measures*

The above results mean that there no loss in generality in writing $B(t-x)/B(t)$ in equation (3.24) as λ_1^{-x}, where λ_1 is the geometric rate of increase of the equilibrium population; in cases where the fitnesses are sensitive to changing demography, an equilibrium can only exist if the population growth-rate is constant, and in cases where the fitnesses are insensitive, the values of $B(t-x)/B(t)$ are unimportant. We can similarly write $k(x, t)$ as independent of t. For notational reasons which will become clearer when we discuss the dynamics of selection in Chapter 4, it is convenient to work with the natural logarithm of λ_1, which we will write as $\hat{r}$ in order to emphasise that it is the intrinsic rate of increase for the equilibrium population (cf. equation (1.45)). The equilibrium fitness measure for equations (3.14) can therefore be written as

$$w_{ij} = \sum_x e^{-\hat{r}x} k_{ij}(x) \tag{3.31}$$

The mean fitness of the equilibrium population is, on this definition, equal to the mean Fisherian reproductive value for a new zygote (cf. equations (1.54)), which is equal to unity. The equilibrium fitness of a genotype can be thought of as its contribution to the reproductive value, at the zygote stage, of the population as a whole.

Similar conclusions concerning the interdependence of demographic stability and genetic equilibrium apply to more general situations. For example, we can use equations (3.20) for the case of non-random mating and differences in demographic parameters between the sexes. Neglecting the terms which are second-order in the gene frequency differences between the sexes, as well as the variance and covariance terms, we obtain

$$p_{ii} = \bar{p}_i^2$$

$$p_{ij} + p_{ji} = 2\bar{p}_i\bar{p}_j$$

Substituting these into equations (3.9) and 3.10), and solving for equilibrium, we find that

$$\hat{p}_i = \sum_x \frac{B(t-x)}{B(t)} \sum_j \hat{p}_j k_{ij}(x, t)$$

$$\hat{p}_i^* = \sum_x \frac{B(t-x)}{B(t)} \sum_j \hat{p}_j k_{ij}^*(x, t)$$

which can be added, to yield

$$1 = \tfrac{1}{2} \sum_x \frac{B(t-x)}{B(t)} \sum_j \hat{p}_j [k_{ij}(x, t) + k_{ij}^*(x, t)] \tag{3.32}$$

This is clearly analogous to equation (3.22a), and can be used to draw the same general conclusions. The equilibrium fitness measure corresponding to that of equation (3.31) is

$$w_{ij} = \tfrac{1}{2} \sum_x e^{-\hat{r}x} [k_{ij}(x) + k_{ij}^*(x)] \tag{3.33}$$

This expression shows that the mean of the fitness measures for the two sexes can, with sufficiently weak selection, be used for determining equilibrium gene frequencies in the same way as the fitnesses when there are no sex differences. The same is true in the corresponding discrete generation case (Wright, 1942; Nagylaki, 1979b).

As discussed by Charlesworth (1972) and Charlesworth and Charlesworth (1973), similar results can be derived for selection models involving such additional factors as recombination and mutation (see also pp. 140–2), and expressions analogous to the discrete generation equations for equilibrium frequencies can always be derived, with the fitness measures of equations (3.31) or (3.33) replacing the corresponding discrete generation formulae.

Finally, we may note that analogous results can be obtained with the continuous-time model; integration replaces summation and the instantaneous logarithmic growth-rate of the equilibrium population replaces $\hat{r}$.

In order to utilise these equilibrium fitness measures, it is obviously usually necessary to know the value of $\hat{r}$. Unless this can be determined in terms of more fundamental genotypic parameters, or is given by empirical data, the equilibrium fitnesses are of uncertain value except when equation (3.25) is satisfied. In the next section, we see how $\hat{r}$ can be determined in the case of a single autosomal locus under selection alone and, in the following section, in the case of selection balanced by mutation.

3.3.3 *Equilibrium in a single-locus system under selection*

Density-independent case. Consider first of all the system described by equations (3.14) with time-independent $k_{ij}(x, t)$ functions. A convenient way of summarising the demographic characteristics of a genotype A_iA_j is by means of the intrinsic rate of increase of a homogeneous population, all of whose members have the reproductive function of A_iA_j, $k_{ij}(x)$. We can call this the intrinsic rate of increase of A_iA_j, denoted by r_{ij}. As in equation (1.45), r_{ij} is the real root of the equation in z

$$w_{ij}(z) = \sum_x e^{-zx} k_{ij}(x) = 1 \qquad (3.34)$$

where $w_{ij}(z)$ generalises the fitness measure of equation (3.31) to arbitrary values of the complex or real variable z.

If we imagine a hypothetical population with *fixed* gene frequencies $p_1, p_2, \ldots p_n$, its intrinsic rate of increase, r_p, can be defined as the real root of

$$\bar{w}(p, z) = \sum_x e^{-zx} \sum_{ij} p_i p_j k_{ij}(x) = 1 \qquad (3.35)$$

Obviously, a real population only attains the rate of growth r_p if the gene frequencies are at equilibrium, but this does not prevent us from using r_p as a useful means of characterising a non-equilibrium population with a given set of gene frequencies.

Using these definitions in the two-allele case, equations (3.22) become

$$1 = \hat{p}_1 w_{11}(\hat{r}) + \hat{p}_2 w_{12}(\hat{r}) \tag{3.36a}$$

$$1 = \hat{p}_1 w_{12}(\hat{r}) + \hat{p}_2 w_{22}(\hat{r}) \tag{3.36b}$$

Noting that $\hat{p}_1 + \hat{p}_2 = 1$, the following pair of homogeneous linear equations in $\hat{p}_1$ and $\hat{p}_2$ can be obtained from these

$$\hat{p}_1[1 - w_{11}(\hat{r})] + \hat{p}_2[1 - w_{12}(\hat{r})] = 0 \tag{3.37a}$$

$$\hat{p}_1[1 - w_{12}(\hat{r})] + \hat{p}_2[1 - w_{22}(\hat{r})] = 0 \tag{3.37b}$$

Solving for $\hat{p}_1$ and $\hat{p}_2$ we obtain the gene-frequency equation for a polymorphic equilibrium

$$\frac{\hat{p}_1}{\hat{p}_2} = \frac{w_{12}(\hat{r}) - w_{22}(\hat{r})}{w_{12}(\hat{r}) - w_{11}(\hat{r})} \tag{3.38}$$

This, as anticipated from the general results derived above, is identical in form with the standard discrete-generation equation for this case (Crow and Kimura, 1970, p. 270). For an equilibrium with $0 < \hat{p}_1 < 1$, we must have either $w_{12} > w_{11}, w_{22}$ or $w_{12} < w_{11}, w_{22}$, excluding the degenerate case $w_{11} = w_{12} = w_{22}$ which implies $r_{11} = r_{12} = r_{22} = \hat{r}$. Furthermore, equations (3.37) imply that $\hat{r}$ must lie between r_{12} and the closer of r_{11} and r_{22}, since the $w_{ij}(z)$ are strictly decreasing functions of z when z is real. This decreasing property also implies that $w_{12} > w_{11}, w_{22}$ is equivalent to $r_{12} > r_{11}, r_{22}$, and that $w_{12} < w_{11}, w_{22}$ is equivalent to $r_{12} < r_{11}, r_{22}$.

To determine $\hat{r}$, we note that, for equations (3.37) to be satisfied for non-zero $\hat{p}_1$ and $\hat{p}_2$, the determinant of their coefficients must vanish, i.e. $\hat{r}$ must satisfy the equation

$$[1 - w_{12}(z)]^2 = [1 - w_{11}(z)][1 - w_{22}(z)] \tag{3.39}$$

This equation was first derived by Norton (1928).

It can be shown as follows that there is only one real value of z which satisfies both this equation and the above requirements for $\hat{r}$. Consider the case $r_{12} > r_{11}, r_{12}$. Then, from the fact that this implies $w_{12} > w_{11}, w_{22}$ at equilibrium and from standard population genetics theory, it follows that the function $\bar{w}(p, \hat{r})$ is at a global maximum with respect to p_1 in the closed interval $[0, 1]$, keeping z

fixed at $\hat{r}$ and using the constraint $\hat{p}_1 + \hat{p}_2 = 1$. Since $\bar{w}(p, z)$ is a decreasing function of z when z is real, this implies $\hat{r} > r_p$ for all $p_1 \neq \hat{p}_1$ in [0, 1]. This excludes the possibility that there is another equilibrium point which is also a maximum of $\bar{w}$ with respect to p_1. The existence of a minimum of $\bar{w}$ is excluded by the fact, also given by standard theory, that this requires $w_{12} < w_{11}, w_{22}$ at the equilibrium point; this is incompatible with $r_{12} > r_{11}, r_{22}$.

We conclude, therefore, that where $r_{12} > r_{11}, r_{22}$, there is a unique equilibrium point with $\hat{r}$ given by equation (3.39), and which corresponds with a global maximum in r_p. A similar argument can be used to establish that $r_{12} < r_{11}, r_{22}$ implies a unique equilibrium with r_p at a global minimum. For practical purposes, $\hat{r}$ must usually be obtained by Newton–Raphson iteration of equation (3.39), given the $k_{ij}(x)$ functions for each genotype. An example of the calculation of $\hat{r}$ and the w_{ij} for a set of genotypes of *Drosophila pseudoobscura* is given by Anderson and Watanabe (1980).

Density-dependent case. A very similar result can be obtained for the case when the population size is regulated by a density-dependent negative feedback mechanism, using the type of model introduced in section 1.4.2 for a genetically homogeneous population (Charlesworth, 1972). It is assumed that we can define a critical age-group consisting of a set S of age-classes, such that the age-specific survival probabilities or fecundities for one or more age-classes are decreasing functions of the numbers of individuals in the critical age-group, either at the present time or at some past time, depending on the mechanism of density regulation. In a population whose numbers are changing in time, the value of $k_{ij}(x, t)$ for a given genotype is controlled by the density of the population; in general, it will depend on some set of numbers of individuals in the critical age-group over a set of earlier times T (equations (1.80) and (1.81)). Writing N_T for this set of numbers, we can replace $k_{ij}(x, t)$ by $k_{ij}(x, N_T)$ in equations (3.14). In a stationary population the numbers in the set N_T are constant in value and equal to the number of individuals in the critical age-group.

The properties of equilibrium populations can be derived if we assume that the net reproduction rate of each genotype, in a population where gene frequencies and population numbers are held fixed, is a strictly decreasing function of the number of individuals in the critical age-group. (This assumption is analogous to the decreasing property of $w_{ij}(z)$ as a function of z, in equation (3.34).) We can define the carrying-capacity, N_{ij}, of A_iA_j as the equilibrium number of individuals in the critical age-group which would be reached by a genetically homogeneous population with the demographic parameters of A_iA_j. This is given as the unique root of the equation

$$w_{ij}(N) = \sum_x k_{ij}(x, N) = 1 \qquad (3.40)$$

where N is a positive real variable corresponding to a fixed number of individuals in the critical age-group. The carrying-capacity, N_p, for a mixed population with fixed gene frequencies $p_1, p_2, \ldots p_n$ can be defined, analogously to r_p in equation (3.35), as the root of

$$\bar{w}(p, N) = \sum_{ijx} p_i p_j k_{ij}(x, N) = 1 \qquad (3.41)$$

Using the decreasing properties of $w_{ij}(N)$ and $\bar{w}(p, N)$ as functions of N, and proceeding in the same way as in the density-independent case, it is possible to establish conditions for the existence of an equilibrium in terms of the N_{ij}; they are identical with the earlier conditions in the r_{ij}. We also have the result that N_p takes a maximum value, $\hat{N}$, at equilibrium when $N_{12} > N_{11}, N_{22}$ and a minimum when $N_{12} < N_{11}, N_{22}$; $\hat{N}$ is given by the analogue of equation (3.39) with N substituted for z. These results are identical with those for the corresponding discrete-generation model (Charlesworth, 1971).

Multiple alleles. With n alleles, the equilibrium is determined by the set of equations corresponding to equations (3.36)

$$\sum_i \hat{p}_i w_{ij}(\hat{r}) = 1 \qquad (3.42a)$$

where $\hat{r}$ is a real root of the determinantal equations corresponding to equation (3.39), with $a_{ij} = 1 - w_{ij}(z)$

$$|A| = 0 \qquad (3.42b)$$

An analogous result holds for the density-dependent case, with the equilibrium carrying-capacity, $\hat{N}$, replacing $\hat{r}$.

Unfortunately, it is not possible to prove the same sort of uniqueness properties for $\hat{N}$ and $\hat{r}$ as in the two-allele case. It can, however, be shown that there is at most one maximum in r_p or N_p with all n alleles present; this corresponds to a stable equilibrium when selection is weak (pp. 156–8). Equation (3.42b) can always be solved numerically, given the $k_{ij}(x)$ functions.

In most other cases, it is not possible to obtain an equation such as (3.39) or (3.42) which determines $\hat{r}$ or $\hat{N}$ in terms of the more basic genotypic parameters. This is so even in the case of a single locus with two alleles when there are demographic differences between the sexes, since the values of the $k_{ij}^*(x, t)$ depend on the composition of the population, so that a detailed specification of this dependence would be necessary to obtain a complete solution for the equilibrium population. This problem can be evaded when selection is weak, using the methods described in section 4.2.1. One important case which *can* be treated satisfactorily, without having to assume weak selection, is when an allele at a locus is maintained at a low frequency as a result of a balance between mutation and selection (Charlesworth and Charlesworth, 1973), and this case will now be considered briefly.

Mutation–selection balance. It is first necessary to define precisely the concept of mutation rate in the context of an age-structured population. For the purpose of discussion, assume a single autosomal locus with wild-type allele A_1 and a deleterious mutant allele A_2. Consider a new A_1A_1 zygote. As this individual develops, its germ-line cells will be exposed to the risk of mutation. Let $u(x)$ be the probability that a germ cell derived from a female aged x contains an A_2 allele in place of one of the A_1 genes inherited from her parents, and let $u^*(x)$ be the corresponding probability for a male. (The chance that both A_1 alleles of an individual experience mutation in the same cell line is negligible.) The *mutation rate*, $u(t)$, for a population at time t which consists entirely of A_1A_1 individuals, is defined as the probability that a randomly chosen gamete involved in a successful fertilisation at time t contains an A_2 allele as a result of mutation. Clearly, we have for $t \geqq d$

$$2B(t)u(t) = \sum_x B(t-x)k_{11}(x, t)u(x)$$

$$+ \sum_x B(t-x)k_{11}^*(x, t)u^*(x) \tag{3.43}$$

This definition can be applied as follows to the case of a population in equilibrium between mutation to A_2 and selection against it. It is reasonable to assume that A_2 is kept rare by selection, so that terms of order $u^2(t)$ and $p_2(t)u(t)$ can be neglected. Equations (3.12) can then be modified in a simple fashion to incorporate the effect of mutations. Carrying out a similar procedure to that involved in deriving equation (3.32), we obtain the following equations for the equilibrium values of the frequencies of A_1 and A_2, averaged over male and female gametes

$$\hat{p} = \sum_x \frac{B(t-x)}{B(t)} \hat{p}_1[\hat{p}_1\bar{k}_{11}(x) + \hat{p}_2\bar{k}_{12}(x)] - u(t) \tag{3.44a}$$

$$\hat{p}_2 = \sum_x \frac{B(t-x)}{B(t)} \hat{p}_2[\hat{p}_1\bar{k}_{12}(x) + \hat{p}_2\bar{k}_{22}(x)] + u(t) \tag{3.44b}$$

where $\bar{k}_{ij}(x) = \frac{1}{2}[k_{ij}(x) + k_{ij}^*(x)]$, and $B(t)$ is now the value for the population as a whole.

For an equilibrium population, there is, as we have seen, no loss in generality in writing $B(t-x)/B(t)$ as $e^{-\hat{r}x}$, where $\hat{r}$ is the rate of growth of the population as a whole. It is easily seen that, in the present case, $\hat{r}$ is equal to r_{11} plus terms of order $u(t)$, so that equations (3.44) can be written, to a satisfactory approximation, as

$$\hat{p}_1 = \hat{p}_1[\hat{p}_1 w_{11} + \hat{p}_2 w_{12}] - u \tag{3.45a}$$

$$\hat{p}_2 = \hat{p}_2[\hat{p}_1 w_{12} + \hat{p}_2 w_{22}] + u \tag{3.45b}$$

where

$$u = \frac{1}{2}\sum_x e^{-r_{11}x}[k_{11}(x)u(x) + k_{11}^*(x)u^*(x)]$$

and

$$w_{ij} = \sum_x e^{-r_{11}x}\bar{k}_{ij}(x)$$

If the relative values of the w_{ij} are written as $1, 1 - hs$ and $1 - s$ for w_{11}, w_{12} and w_{22} respectively, equations (3.45) can be re-arranged to yield the equilibrium frequency of A_2 in the standard form employed in discrete-generation models (Crow and Kimura, 1970, p. 266)

$$s(1 - 2h)\hat{p}_2^2 + hs\hat{p}_2 - u = 0 \qquad (3.46)$$

This equation has the familiar approximate solution for the dominant or semi-dominant case

$$\hat{p}_2 \cong u/hs \quad (h > 0) \qquad (3.47a)$$

and, for the recessive case,

$$\hat{p}_2 \cong \sqrt{u/s} \quad (h \cong 0) \qquad (3.47b)$$

Analogous results can be derived for the density-dependent case, with the fitnesses all being evaluated at $N = N_{11}$. It may be noted that equation $(3.47a)$ is valid for the case of non-random mating with respect to age and demographic differences between the sexes regardless of the intensity of selection, since second-order terms in $\hat{p}_2^2$ are neglected in obtaining it and the terms in $\bar{\delta}_1^2$ and V in equations (3.21) are also of order $\hat{p}^2$ in this case. A similar treatment can also be made for the case of a sex-linked locus.

3.4 Biological applications of the results

This section surveys the implications of the theoretical results derived in this chapter for experimental and observational studies of natural populations, with emphasis on those properties which are peculiar to age-structured as opposed to discrete-generation populations.

Probably the most important single respect in which age-structured populations differ from those with discrete generations is the interdependence of demographic stability and genetic equilibrium. As we saw in section 3.3.1, we generally expect an age-structured population to remain precisely in genetic equilibrium only if the population is demographically stable. Changes in the age-structure of the population, induced by ecological changes, may therefore cause changes in allele frequencies at loci under selection, even if the force of selection is unchanged at the fundamental level of age-specific survival probabilities and fecundities, i.e. if the relative values of the $k_{ij}(x)$ functions for different genotypes remain unchanged for each value of x.

The possible magnitudes of such shifts in gene frequencies, and their relationship to the nature and strength of selection, were investigated by Charlesworth and Giesel (1972a) by means of computer calculations of population trajectories of two-allele systems, using equations (3.14). They showed that cyclical fluctuations in gene frequencies could be generated in populations exposed to regular cycles of population growth, imposed by an external ecological factor which varied in intensity in time. An example of such an effect is shown in Figure 3.1. The fluctuations in population size were produced by periodic switches of the $k_{ij}(x)$ functions such that, at a point of change over from one regime to a new one, the old $k_{ij}(x)$ values for each genotype and age were multiplied by the same factor, which was greater or less than one depending on the desired direction of change in population growth-rate. This corresponds to the effect of an environmental factor causing periodic changes in the fecundities, and which affects each genotype and age equally. A_1A_1 initiated and completed reproduction later than A_2A_2. The heterozygote A_1A_2 combined the favourable characteristics of the homozygotes, thereby ensuring maintenance of the polymorphism in each environment (section 4.3.1). In this example, there were three environmental states, following each other in regular succession and corresponding to the equilibrium states shown in Table

Figure 3.1. Fluctuations in gene frequency at a polymorphic locus, induced by cyclical changes in the population growth-rate. The solid line shows the successive values of log $B(t)$, and the dashed line the frequency of allele A_1 of the system shown in Table 3.2. Each genotype had 13 age-classes. (After Charlesworth and Giesel, 1972, Fig. 2.)

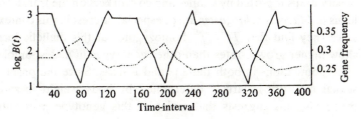

3.2, which would be attained if the population were left in a given state for a sufficiently long time.

Table 3.2. *Equilibria corresponding to different environments*

Environment	$\hat{r}$	$\hat{p}_1$	w_{11}/w_{12}	w_{22}/w_{12}
1	−0.10	0.468	0.886	0.898
2	0.00	0.275	0.816	0.930
3	0.10	0.149	0.743	0.955

As can be seen from Figure 3.1, the actual time-course of change of the population is such that the gene frequency and growth-rate fluctuate within the two extreme equilibrium values; when the population is switched into a new state, it starts moving towards the corresponding equilibrium, but does not have time to reach it before the environment changes again. The reason that allele A_1 is favoured when the population is declining but A_2 is favoured in the period of population increase, is that the A_2A_2 homozygotes reproduce relatively early. The weighting factor $\exp(-\hat{r}x)$ in the equilibrium fitness in equation (3.31) gives increased weight to early as opposed to late reproduction when $\hat{r}$ is positive, and vice versa when $\hat{r}$ is negative. From equation (3.38), one would expect, therefore, A_1 to be most frequent in populations with low $\hat{r}$, since the relative values of the $k_{ij}(x)$ are independent of the environmental state in this model.

This type of model is of interest in relation to observations on natural populations, where fluctuations in allele frequencies have often been reported (e.g. Dobzhansky, 1943; Dubinin and Tiniakov, 1945; Gershenson, 1945; Tamarin and Krebs, 1969; Gaines and Krebs, 1971; Gaines, McCleaghan and Rose, 1978). The observations reported by Gaines and co-workers on the transferrin locus in *Microtus ochrogaster* are of especial interest in this context, since they find that Tf^E/Tf^E homozygotes at this diallelic locus show higher growth-rates than other genotypes in males sampled from populations in both Indiana and Kansas. Since the onset of sexual maturity is largely determined by body weight in small mammals, this suggests that males of this genotype may enter reproductive life earlier than the others. This is a species which shows population cycling, and there is evidence that the Tf^E allele tends to increase in frequency during the phase of population increase.

Although it is not possible in such cases of gene frequency fluctuations to exclude other interpretations, such as selective responses to environmental factors like temperature or to population density itself (cf. King and Anderson, 1971; Charlesworth and Giesel, 1972*b*), nevertheless investigators should bear in mind the possibility that gene-frequency changes may be due to a purely mechanical shift in genotypic fitness as a result of changing demography, and are not necessarily caused by changes in the selection regime at the level of age-specific survival probabilities and fecundities.

The idea that genotypic fitnesses in an age-structured population may depend on the overall demographic structure of the population is not a new one. For instance, Bodmer (1968) suggested that the fitnesses of sufferers from genetic diseases with a late age of onset may change in response to changes in the mean age of reproduction of the whole population. The selection models and resulting expressions for equilibrium fitness described in the present chapter provide, however, an exact genetic foundation for analysing the effects of demography on fitnesses, as will be seen in section 3.4.1.

3.4.1 *Factors influencing relative fitnesses*

The way in which demographic factors can influence relative genotypic fitnesses, and thereby alter the genetic composition of equilibrium populations, can be understood in terms of the equations for equilibrium genotypic fitnesses, such as (3.31) and (3.33). For simplicity, the former is used here as a basis of discussion. Let us first consider the effect of a change in an environmental source of mortality, which affects the $l(x)$ functions of the various genotypes in a similar way. If $l_s(x)$ is the probability of survival to age x for a standard genotype, we can write

$$l_{ij}(x) = l_s(x)l'_{ij}(x)$$

where $l'_{ij}(x)$ is a function which characterises A_iA_j, and $l_s(x)$ is common to all genotypes. Changes in the level of an external source of mortality which affects each genotype equally can be absorbed into $l_s(x)$. For an equilibrium population in stable age-distribution, the fitness of A_iA_j is given by

$$w_{ij} = \sum_x e^{-\hat{r}x} l_s(x) l'_{ij}(x) m_{ij}(x) \qquad (3.48)$$

If the population experiences a change in the level of mortality, there may be a change in its overall age-distribution when equilibrium is re-established. This would happen if, for instance, the population was held in check by density-dependent factors affecting the survival of immature stages, and the survival probability of each adult age-class were altered ($\hat{r}$ in this case would be zero for the old and new equilibrium). Any such change will normally be expressed as a shift in $e^{-\hat{r}x} l_s(x)$ as a function of age (cf. section 1.3.2). If the age-structure were shifted in favour of younger reproductive individuals, $e^{-\hat{r}x} l_s(x)$ must be changed in such a way as to give greater weight to the $l'_{ij}(x) m_{ij}(x)$ for younger age-classes in equation (3.48). Genotypes which have relatively high $k_{ij}(x)$ functions for younger ages, such as genotypes which initiate reproduction early in life or attain their peak fecundity early, would tend to have higher relative fitnesses in the new equilibrium population. The opposite would obviously be true if the age-structure of the population became weighted in favour of older individuals due to a relaxation in mortality.

A very similar analysis can be carried out for the case of a change in an environmental factor affecting age-specific fecundities, as in the computer model described above. Here, we can write $m_s(x)$ for the fecundity function of the standard genotype, and $m'_{ij}(x)$ for the fecundity function characterising A_iA_j, so that

$$w_{ij} = \sum_x e^{-\hat{r}x} l_{ij}(x) m_s(x) m'_{ij}(x) \qquad (3.49)$$

If the population is density-independent, an increase in $m_s(x)$ at each age results in an increase in $\hat{r}$, with a shift in the overall age-structure towards younger individuals which is reflected in $e^{-\hat{r}x} l_s(x)$, and conversely if $m_s(x)$ is reduced. A change in the age-structure is therefore reflected in a change in the weighting of reproduction with respect to age.

On the other hand, changes in mortality or fecundity which do not result in any changes in the age-structure of the reproductive age-classes fail to produce alterations in the weighting of reproduction at different ages in equations (3.48) or (3.49), unless the changes are directed specifically at certain age-groups. For instance, if the same extrinsic probability of death were added to each age-class of each genotype, its effect on age-structure in a density-independent environment would be exactly cancelled out by a change in $\hat{r}$ (cf. Chapter 1, pp. 34–5). There would be no change in $e^{-\hat{r}x}l_s(x)$ and hence no change in the weighting of reproduction in equation (3.48). Similarly, in a population which is regulated by density-dependent mortality of immature stages, an increase in $m_s(x)$ for each age would be compensated for by an increased mortality of the immature individuals, resulting in no change in $\hat{r}$, no change in the age-structure of reproductively active stages, and no increased weighting of early reproduction. Only if the increase in $m_s(x)$ were concentrated in early or late reproductive age-classes would there be a change in weighting without a change in the age-structure of the reproductive part of the population.

These considerations suggest that demographic changes which affect the age-structure of the reproductively active part of the population can cause shifts in gene frequencies at loci under selective control, provided that the relative values of the $k_{ij}(x)$ functions for different genotypes vary sufficiently with age. Conversely, demographic changes which leave the age-structure of the reproductive age-classes unchanged are unlikely to cause gene-frequency changes, unless they are due to a source of mortality or a factor affecting fecundity which is directed specifically at certain genotypes or at a specific group of ages of the reproductive individuals. Such demographically induced changes in relative fitnesses and gene frequencies cannot occur to any significant extent if the $k_{ij}(x)$ functions are close to satisfying equation (3.25), i.e. if selection is non-specific with respect to age. As shown by Charlesworth and Giesel (1972a), selective situations in which the shapes of the $m_{ij}(x)$ functions with respect to age vary between genotypes are highly favourable for this effect. Mortality differences between genotypes are unlikely to result in shifts in relative fitnesses with demographic changes, unless there are pronounced genotypic differences in the patterns with respect to age of the survival probabilities from one age-class to the next, among the reproductively active age-classes. Such a case occurs, for example, if one genotype is afflicted with a reduced survival probability, but the onset of this is delayed until well after the beginning of reproduction.

Shifts in gene frequencies due to demographic changes might well be expected in populations which have colonised a new habitat where density-dependent restraints are lifted and there is a high population growth-rate. Although one would probably expect only a limited number of loci to be under sufficiently powerful selective control of a suitable sort for such changes to be induced, hitch-hiking effects on closely-linked loci (Maynard Smith and Haigh, 1974) might contribute to an overall disturbance in the gene pool, and add an additional element to the 'genetic revolution' which Mayr (1963, pp. 526–61) suggested could take place in a small isolate.

3.4.2 *The measurement of fitness in human populations*

The results derived in this chapter are of obvious relevance to human geneticists interested in measuring genotypic fitnesses, for example in connection with the 'indirect' method of estimating mutation rates from the theoretical formulae for the balance between mutation and selection (equation (3.47)). For such purposes, it is usually assumed that the population is at equilibrium, so that the fitness measures described in this chapter are appropriate. Since human populations generally show marked differences in vital statistics between males and females, the more general fitness measure of equation (3.33) is preferable, assuming that the necessary information on the vital statistics of the genotypes involved is available. It should be borne in mind that this formula is strictly valid only when the effects of non-random mating with respect to age, and of sex differences in gene frequencies, can be treated as second-order (section 3.2.4). As pointed out in section 4.3.1, this is necessarily the case for a rare non-recessive gene, so that studies of selection on rare dominant or semi-dominant alleles can be carried out without serious error, using equation (3.33). For rare recessives, or polymorphic loci, the use of the formula requires relatively small selection intensities.

The greatest difficulty in using the formula is that detailed information on the vital statistics of the genotypes concerned is rarely available. Cavalli-Sforza and Bodmer (1971, Chapter 6) review various indirect methods of obtaining vital statistics for genetic traits. Some examples of the applications of these methods in connection with the equilibrium fitness measure described here are given by Charlesworth and Charlesworth (1973). We shall briefly consider some of their findings.

One of the few human genetic traits for which a reasonably complete life-table has been estimated directly is the Marfan syndrome, a disease of the connective tissue and vascular system which is controlled by a rare dominant gene. Figure 3.2 shows the life-tables obtained by Murdoch *et al.* (1972) for U.S. men and women suffering from this disorder, together with normal life-tables for the same period of time. It will be seen that the disease first manifests itself as an increased probability of death early in reproductive life; the mean age of death is 32. No information on the age-specific fecundities of sufferers from the syndrome appears to be available, so that we cannot find their fitness directly; but an over-estimate can be obtained by assigning them the same $m(x)$ and

Figure 3.2. Survival functions $l(x)$ and $l^*(x)$, at intervals of one year, for individuals with Marfan syndrome (⊙——⊙, females; ▲——▲, males) and for average individuals (●——●, females; ★——★, males) from the same (U.S.) population. (After Murdoch *et al.*, 1972, Fig. 2.)

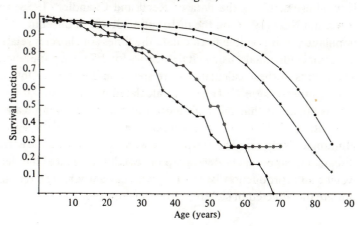

$m^*(x)$ functions as normal individuals. Using census data for the U.S. population of 1964, and the $\hat{r}$ value corresponding to the female data for that year, we find that $\sum e^{-\hat{r}x}k(x)$ for the Marfan syndrome is 0.91, and $\sum e^{-\hat{r}x}k^*(x) = 0.78$, giving an overall fitness value of 0.85. The relatively small selection coefficient of 0.15 against this trait, despite its substantial effect on the mortality of adults, is due to its delayed age of onset. Nevertheless, fitness is relatively insensitive to changing demography in this case. If the vital statistics for the 1939–41 U.S. population are used as a basis for calculation, the fitness value is almost unchanged, despite the fact that, because of lower fecundity, this population has an intrinsic rate of increase of -0.002 per year compared with $+0.016$ for 1964. This illustrates the conclusion reached above concerning the need for strong differences between the mortality patterns with respect to age of different genotypes during reproductive life if demographic structure is to affect relative fitnesses. The time of onset of the mortality effect of the Marfan syndrome coincides approximately with the onset of reproduction, and so this condition is not satisfied.

This insensitivity to demographic change is not always the case, however. Charlesworth and Charlesworth (1973) examined the fitnesses of carriers of Huntington's chorea, another rare dominant disease. This is a disease with a late and variable age of onset. Cavalli-Sforza and Bodmer (1971) suggested that a useful approximate estimate of the effect of the disease on fitness could be obtained by assuming that sufferers cease to reproduce as soon as it is manifested. Using the data of Reed and Chandler (1958) and Reed and Neel (1959) on the distribution of age of onset in a U.S. population, it is possible to calculate the fitness of choreics relative to normal individuals, using vital statistics for different populations. The results of these calculations are shown in Table 3.3. Only the female fitnesses, $\sum e^{-\hat{r}x}k(x)$, were calculated in this case.

It is apparent that the high-mortality, rapidly-growing Taiwan population of 1906 has the lowest selection coefficient against Huntington's chorea. This is what one would expect from the earlier discussion, since such demographic conditions place the least weight on reproduction late in life, whereas early reproduction is not greatly affected in sufferers from this disease.

Table 3.3. *Selection coefficients against Huntington's chorea in populations with different demographic conditions*

Population	Intrinsic rate of increase (per year)	Selection coefficient against chorea
U.S. (1939–41)	−0.002	0.152
U.S. (1964)	+0.016	0.126
Taiwan (1906)	+0.021	0.088

There are numerous other diseases in man which either have a simple mode of genetic determination, or for which there is evidence of the involvement of genetic factors, and which do not manifest themselves until after the start of reproductive life. Examples of the former include genetically controlled cancers such as multiple polyposis of the colon (Reed and Neel, 1955) and of the latter, such traits as high blood pressure (Cavalli-Sforza and Bodmer, 1971, pp. 583–5). In all these cases, we may expect that the reduction in overall mortality rates which has accompanied advancing civilisation will have increased the selection pressure against the genes that are implicated in such conditions. Their frequencies in contemporary populations may therefore be far from reflecting the equilibrium which they would eventually reach if demographic change were to cease. The estimation of mutation rates by the indirect method is, accordingly, especially dangerous for genes affecting characters of this sort.

Approximate estimators of fitness. In most cases, even such indirect approaches to determining genotypic vital statistics are not possible, and various approximations have to be resorted to. Provided that the population is not growing too fast, the mean completed family size, discounting individuals who die before reproductive maturity, should provide a reasonably good approximation. If all that is known is that a mutant phenotype suffers an enhanced average probability of death of Q per age-class, without any information on the relationship of Q to age, it is possible to approximate w_{ij} in equation (3.33) by

$$\tfrac{1}{2}\sum e^{-\hat{f}x}(1-Q)^x[k(x)+k^*(x)] \cong 1-\tfrac{1}{2}Q\sum x\, e^{-\hat{f}x}[k(x)+k^*(x)]$$

where $k(x)$ and $k^*(x)$ are given by the vital statistics for the normal population. The multiplier of Q on the right-hand side of this expression is the mean generation-time of the population (cf. equation (1.47c)). Similarly, if all that is known is that the mutant has a fecundity of $1-K$ relative to normal, the best estimate of its fitness is $1-K$.

Charlesworth and Charlesworth (1973) have discussed various other measures of genotypic fitnesses which have been proposed by other authors, and compared them with the values given by equation (3.31), for the case of Huntington's chorea. The measure discussed here has the definite advantage of being based on a clear-cut genetic model, whereas earlier measures rest on somewhat intuitive grounds. Provided that adequate data are available, there is no greater computational difficulty in using equations (3.31) or (3.33) than the less well-founded measures, with modern computing methods. Of course, the inherent sampling error and biases in the available data will usually mean that the use of an incorrect fitness measure is only one of many sources of error in estimation.

4

Selection: dynamic aspects

4.1   Introduction

This chapter is concerned with the analysis of the process of gene frequency change under selection in age-structured populations, in particular with the question: to what extent can this process be understood in terms of fixed parameters which characterise individual genotypes, analogous to the fitnesses of genotypes in discrete-generation models? The first investigation of the problem of the dynamics of selection in the context of age-structured populations was that of Norton (1928). (According to Haldane (1927a), Norton started work on the problem as early as 1910 and had established his results by 1922.) He studied the case of a single diallelic locus with time-independent and density-independent demographic parameters for each genotype, with no sex-differences, using the continuous-time equivalents of equations (3.14). He showed that the long-term results of selection (whether one of a pair of alleles is fixed or a polymorphism is maintained) are completely predicted by the relationships between the intrinsic rates of increase of the three genotypes, in the same way that the outcome of selection can be predicted from the genotypic fitnesses in the discrete-generation case. Haldane (1927a) established some approximate results for the rates of spread of rare dominant and recessive genes in slowly growing populations, and concluded that selection proceeded in much the same way as in the discrete-generation case. Both Haldane and Norton based their work on equations which explicitly incorporated the characteristics of age-structured populations, in contrast to the approach of Fisher (1930, 1941) who used differential equations to describe gene frequency change with continuous-time models but did not derive these from more basic equations.

Later work (Haldane, 1962; Pollak and Kempthorne, 1971; Charlesworth, 1973, 1974a, 1976) has established that no single

153

parameter can be regarded as the fitness of a genotype with arbitrary selection intensities. When selection is weak, however, it is possible to show that the intrinsic rate of increase of a genotype can be equated with its fitness, with time-independent and density-independent demographic parameters (Charlesworth, 1974a, 1976). Useful results can also be obtained for the density-dependent case with weak selection.

In this chapter, we start by examining how the basic equations for gene frequency and population size, such as equations (3.12) and (3.14), can be approximated with weak selection to yield equations which are similar in form to those for the usual discrete-generation models. We then go on to consider the case of arbitrary selection intensities, by means of the local stability properties of equilibria. The probability of survival of a favourable mutant gene is also treated. The problem of the long-term effects on gene frequencies of a given selective regime is discussed. Finally, the prediction of the response to selection on a quantitative character is considered.

4.2   Approximate equations with weak selection

Most of the models in this section are developed in terms of a single autosomal locus with an arbitrary number, n, of alleles, as in the preceding chapter, although the case of two loci is considered briefly. In addition, discussion is largely confined to the case of equations (3.14), with some mention of the problem of sex differences in demographic parameters. Both density-independent populations (with time-independent $k_{ij}(x)$ functions) and density-dependent populations are considered. Before proceeding to the detailed results, it is useful to have a clear definition of what is meant by weak selection. We choose some genotype as an arbitrary standard, and write its reproductive function as $k_s(x, t)$. The reproductive function for another genotype A_iA_j can be written as $k_{ij}(x, t) = k_s(x, t) + \varepsilon_{ij}(x, t)$, where $|\varepsilon_{ij}(x, t)| < \varepsilon$ for all i, j and x; ε is a positive number which measures the strength of selection. Selection is said to be weak when terms $O(\varepsilon^2)$ can be neglected, compared with terms $O(\varepsilon)$.†

---

† As is conventional in mathematical usage, a function of ε, $f(\varepsilon)$, is said to be of order ε or $O(\varepsilon)$ if $|f(\varepsilon)/\varepsilon| < \alpha$ when ε is taken sufficiently small, where α is a constant independent of ε. If $|f(\varepsilon)/\varepsilon|$ approaches 0 as $\varepsilon \to 0$, we write $f(\varepsilon) = o(\varepsilon)$.

4.2.1 *The density-independent case*

Approximate equations for $B(t)$. The first step in obtaining expressions for changes in gene frequency and population growth-rate is to obtain an approximation for $B(t)$. From equation (3.14*b*) we have the exact expression for $t \geq d$

$$B(t) = \sum_x B(t-x) \sum_{ij} p_i(t-x)p_j(t-x)k_{ij}(x) \tag{4.1}$$

In constructing an approximate solution to this equation, we need the result that the change in gene frequency between times t and $t+1$ (more technically, the *first difference* in gene frequency), which we can write as $\Delta p_i(t)$ for allele A_i, is of order ε for sufficiently large t. This result, which is what one intuitively expects, is derived in Appendix 2. As mentioned there, convergence to terms of order ε should normally be fairly rapid.

We now choose a time t_0 which is sufficiently large that convergence of the Δp_i has taken place for each allele, and consider a time t such that t_0 is previous to $t-d$. We can write $p_i(t-x) = p_i(t_0) + \zeta_i(t-x)$, where ζ_i is $O(\varepsilon)$ if t and t_0 are sufficiently close, since ζ_i is simply the sum of the Δp_i between t_0 and $t-x-1$. Equation (4.1) can therefore be rewritten as

$$B(t) = \sum_x B(t-x) \left[\sum_{ij} p_i(t_0)p_j(t_0)k_{ij}(x) \right.$$

$$\left. + 2\sum_{ij} \zeta_i(t-x)p_j(t_0)k_{ij}(x) \right] + O(\varepsilon^2) \tag{4.2}$$

But $\sum_j p_j(t_0)k_{ij}(x) = \sum_{ij} p_i(t_0)p_j(t_0)k_{ij}(x) + O(\varepsilon)$. Substituting this into equation (4.2), and using the relation $\sum_i \zeta_i = 0$, we obtain

$$B(t) = \sum_x B(t-x) \sum_{ij} p_i(t_0)p_j(t_0)[k_{ij}(x) + O(\varepsilon^2)] \tag{4.3}$$

Applying to this equation the type of argument used in Appendix 2, we obtain $\Delta \ln B \sim r_0 + O(\varepsilon^2)$, so that

$$B(t-x) \sim B(t) e^{-r_0 x} + O(\varepsilon^2) \tag{4.4}$$

where r_0 is the intrinsic rate of increase associated with the mean reproductive function for time t_0, $\sum_{ij} p_i(t_0)p_j(t_0)k_{ij}(x)$. In other words, r_0 satisfies the equation

$$\sum_x e^{-zx} \sum_{ij} p_i(t_0) p_j(t_0) k_{ij}(x) = 1$$

If terms $O(\varepsilon^2)$ are neglected, this gives us the approximate asymptotic result which we need. This can validly be done if selection is sufficiently weak (ε is sufficiently small), so that second-order terms can be neglected in the total gene frequency change which takes place over the period of time needed to stabilise the age-structure of a population with a constant reproductive function. In most cases of biological interest this should not be a long time, so that the condition is not unduly restrictive.

Intuitively, this result can be viewed in the following way. Equation (4.3) shows that the effect of selection on the mean reproductive function of the population is second order compared with the selection intensity, so that the population can be treated as though it has a fixed reproductive function and hence a fixed rate of population growth over a sufficiently short period of time.† This can be compared with the result for discrete-generation models that the rate of change of population mean fitness is second-order, with respect to the fitness differences between genotypes (Crow and Kimura, 1970, pp. 208–9).

Approximate equations for Δp_i: type 1. From equation (3.14a) we have the gene frequency equation for $t \geqq d$

$$B(t) p_i(t) = \sum_x B(t-x) p_i(t-x) \sum_j p_j(t-x) k_{ij}(x) \qquad (4.5)$$

---

† Of course, as time passes, $t - t_0$ will become so large that the $O(\varepsilon^2)$ terms in equation (4.4) can no longer be neglected. This problem can be overcome by shifting to a later time t_1, in place of t_0, to which there corresponds a different intrinsic rate of increase r_1 (cf. p. 157). Provided that ε is sufficiently small, and the demography of the population is such that the time needed to stabilise age-structure is small compared with the time needed for a change in gene frequency of $O(\varepsilon)$, there will always be *some* intrinsic rate of increase which can be substituted into equation (4.4).

Substituting from equation (4.4), this becomes (for t sufficiently close to t_0)

$$p_i(t) = \sum_x e^{-r_0 x} p_i(t-x) \sum_j p_j(t-x) k_{ij}(x) + O(\varepsilon^2) \qquad (4.6)$$

This equation can be simplified further if we approximate $p_i(t-x)$ by $p_i(t) - x\Delta p_i(t)$. This type of approximation was first used in the context of models of age-structured populations by Haldane (1927a, 1962). It is valid under conditions discussed in Appendix 2. Substituting into equation (4.6), and dropping the arguments from $p_i(t)$ and $\Delta p_i(t)$, we obtain

$$p_i = p_i \sum_x e^{-r_0 x} \sum_j p_j k_{ij}(x)$$

$$-\Delta p_i \sum_x x\, e^{-r_0 x} \sum_j p_j k_{ij}(x) + O(\varepsilon^2) \qquad (4.7)$$

In order to proceed further with this equation it is necessary to eliminate the unknown quantity r_0, which depends on the arbitrary choice of time t_0. This can be done by again using the argument which led from equations (4.1) to (4.3); we have

$$\sum_{ij} p_i(t_0) p_j(t_0) k_{ij}(x) = \sum_{ij} p_i p_j k_{ij}(x) + O(\varepsilon^2)$$

This means that, to order ε^2, r_0 can be replaced in equation (4.7) by r_p as defined by equation (3.35); the subscript p represents the set of gene frequencies at time t.

It is useful to define the following quantities;

$$T_{ij}(z) = \sum_x x\, e^{-zx} k_{ij}(x) \qquad (4.8a)$$

$$T(z) = \sum_x x\, e^{-zx} \sum_{ij} p_i p_j k_{ij}(x) \qquad (4.8b)$$

$$w_i(z) = \sum_j p_j w_{ij}(z) \qquad (4.8c)$$

where the $w_{ij}(z)$ are given by equation (3.34).

Using these definitions, and replacing r_0 by r_p, equation (4.7) can be rearranged to give the first type of approximate expression for Δp_i

$$T(r_p)\Delta p_i = p_i[w_i(r_p) - 1] + O(\varepsilon^2) \tag{4.9}$$

$T(r_p)$ in this equation may be regarded as a measure of the generation time of the population at time t (cf. Chapter 1, p. 33), and provides a time-scale for the process of gene frequency change.

This equation is closely connected with the equilibrium theory developed in section 3.3.2. When $r_p = \hat{r}$, as given by equations (3.42), we have $w_i = 1$ for all alleles which are present in the population. Hence, the equilibrium solution of equation (4.9) is, if we neglect the $O(\varepsilon^2)$ terms, identical with that given by the exact equations of Chapter 3. Furthermore, standard theory tells us that, as in the two-allele case discussed on pp. 136–8, this equilibrium corresponds to a stationary point in the function $\bar{w}(p, \hat{r})$ which was defined by equation (3.35), with respect to variation in gene frequency while holding r_p constant at $\hat{r}$ (cf. Crow and Kimura, 1970, pp. 272–7). This implies that, as in the two-allele case, r_p is at a stationary point at the equilibrium. In the neighbourhood of equilibrium, equations (4.9) can thus be satisfactorily approximated by substituting $\hat{r}$ for r_p. It follows that the $w_{ij}(\hat{r})$ control the local stability of the equilibrium in the same way as the fitnesses W_{ij} in the discrete-generation case. From the standard theory of the discrete-generation case, the mean fitness of the population is at a global maximum at a stable equilibrium point with all alleles present, and there is only one such equilibrium. Comparing the discrete-generation case with the present system, using the $w_{ij}(\hat{r})$ as fitness parameters, we see that $\bar{w}(p, \hat{r})$ is at a global maximum at a stable equilibrium point, neglecting the $O(\varepsilon^2)$ terms; it follows that $\hat{r}$ is a maximum of r_p at such a point, and that there can be at most one stable equilibrium with all alleles present. In the case of a diallelic locus, it follows that the equilibrium is locally stable if $r_{12} > r_{11}, r_{22}$, and is unstable if $r_{12} < r_{11}, r_{22}$. Study of equation (4.9) shows that these relations also determine the global stability conditions with two alleles and weak selection (cf. Charlesworth, 1974a).

Approximate equations for Δp_i: type 2. Although equation (4.9) provides a useful characterisation of the dynamics of the system, it suffers from the weakness that r_p is itself an implicit function of the gene frequencies, so that it is tedious to calculate the rate of change of gene frequencies at any given point. We now develop a further approximation to this equation, which yields an expression for Δp_i that depends only on gene frequency and on the genotypic intrinsic rates of increase.

We note that we can approximate $w_{ij}(r_p)$ (given by equation (3.34)), using Taylor's theorem. We have

$$1 = w_{ij}(r_{ij}) = w_{ij}(r_p) + (r_{ij} - r_p)(\partial w_{ij}/\partial z)_{r_p} + O(\varepsilon^2)$$

so that

$$w_{ij}(r_p) - 1 = (r_{ij} - r_p)T_{ij}(r_p) + O(\varepsilon^2) \tag{4.10}$$

Substituting this into equation (4.9), and noting that $T_{ij}(r_p) = T(r_p) + O(\varepsilon)$, we obtain the final result

$$\Delta p_i = p_i(r_i - \bar{r}) + O(\varepsilon^2) \tag{4.11}$$

where

$$r_i = \sum_j p_j r_{ij} \quad \text{and} \quad \bar{r} = \sum_{ij} p_i p_j r_{ij}.$$

An analogous procedure can be gone through for the continuous-time model, for which we obtain

$$\frac{dp_i}{dt} = p_i(r_i - \bar{r}) + O(\varepsilon^2) \tag{4.12}$$

where the genotypic intrinsic rates of increase are now defined by the integral equation analogue of equation (3.34).

It is clear from these results that, when selection is sufficiently weak, the use of the genotypic intrinsic rates of increase provides a useful approximate method of calculating rates of gene frequency change in a density-independent environment, with the r_{ij} playing much the same role as the W_{ij} in the discrete-generation case with weak selection (in which case the normalising term $\bar{W}$ on the right-hand side of equations (3.15) can be dropped; see Crow and

Kimura, 1970, p. 191). Equation (4.12) with weak selection is also similar in form to Fisher's Malthusian parameter equation for gene frequency change (Fisher, 1930, 1941). If selection is too strong, however, the $O(\varepsilon^2)$ terms in equations (4.11) and (4.12) may become significant, and one cannot expect the main term to provide an adequate approximation.

Rate of change of the population growth-rate. The similarity between equation (4.11) and the more usual equations for gene frequency change suggests that it should be possible to derive an expression for the rate of change of the population growth-rate, analogous to that given by Fisher's Fundamental Theorem of natural selection. This can be done as follows. Ignore for the moment the $O(\varepsilon^2)$ terms in the equations. The results given above show that the rate of growth of a population at time t is approximately r_p. Using equation (4.10) it is easy to see that r_p can be approximated to $O(\varepsilon^2)$ by $\bar{r}$, so that we have (approximately)

$$\Delta^2 \ln B(t) = \Delta \bar{r} \tag{4.13}$$

By standard theory (Crow and Kimura, 1970, pp. 205–10), we have

$$\Delta \bar{r} = 2 \sum_i p_i (r_i - \bar{r})^2 + O(\varepsilon^3) \tag{4.14}$$

where the term on the right-hand side is the additive genetic variance in the intrinsic rate of increase. Putting equations (4.13) and (4.14) together gives us a version of Fisher's Fundamental Theorem: the rate of change of the population growth-rate is approximately equal to the additive genetic variance in the intrinsic rate of increase.

This result as it stands is purely formal, however, since terms in ε^2 were neglected before taking the differences in equation (4.13); if the first differences in these terms were themselves of order ε^2, they would contribute significantly to the final result and destroy the approximation. Retracing our steps through the successive approximations, it is possible to show, by arguments similar to those of Appendix 2, that the first differences in the $O(\varepsilon^2)$ terms are themselves $O(\varepsilon^3)$, so that equations (4.13) and (4.14) combined give

$$\Delta^2 \ln B(t) = 2 \sum_i p_i (r_i - \bar{r})^2 + O(\varepsilon^3) \qquad (4.15)$$

Accuracy of the approximations.   The utility of the results derived above depends on being able to neglect the $O(\varepsilon^2)$ and $O(\varepsilon^3)$ terms in the equations for gene frequency change and change in population growth-rate. It is obviously important to determine in practice how weak selection must be for this to be done without significant error. This question was investigated numerically by Charlesworth (1974*a*), who compared the results obtained from exact population trajectories with the approximate values of Δp_i and $\Delta^2 \ln B$ obtained from equations (4.9), (4.11) and (4.15). He used a ten age-class system with the T_{ij} of equation (4.8*a*) all approximately equal to 4. A crude measure of the strength of selection in per generation terms is therefore obtainable by multiplying the differences in genotypic intrinsic rates of increase by 4. Cases of dominant, recessive and semi-dominant favourable alleles in diallelic systems, as well as heterozygote superiority, were studied. Even with selective differences of up to 40% per generation, there is fairly good agreement between the exact and approximate values for Δp_i and $\Delta^2 \ln B$, after a few generations have elapsed and the age-structure of the population has stabilised. As might be expected from its relationship to the exact equilibrium equations, equation (4.9) seems to provide a better approximation than (4.11) for the neighbourhood of a polymorphic equilibrium. Away from such a point, equation (4.11) provides, if anything, a better fit. Since it does not depend on any parameters other than the gene frequencies and the r_{ij}, it is to be preferred to equation (4.9). In using these approximate equations, it should be noted that they are designed to predict the rates of change of gene frequencies and population growth-rate, given the current gene frequencies. It cannot be expected that they will necessarily provide an accurate basis for calculating a whole population trajectory, since the errors will tend to accumulate over successive time-intervals.

Sex differences in demographic parameters. The problem of handling sex differences in demographic parameters, which was avoided in the above treatment by basing it on equations (3.14), can be approached in the following way when there is random mating with respect to age (cf. Charlesworth, 1974a). Consider an arbitrary standard and homogeneous population with female reproductive function $k_s(x)$ and growing at its intrinsic rate of increase r. Each male genotype can be characterised by its age-specific fecundity function when introduced at a very low frequency into this population; from the assumption of a stable age-distribution, these fecundities will be independent of time. Denote the 'standard' reproductive function for A_iA_j males as given by this procedure by $k_{ij}^*(x)$, with a corresponding intrinsic rate of increase r_{ij}^*. In a real population, with a mixture of genotypes changing in frequency under selection, the $k_{ij}^*(x, t)$ functions at a given time, t, must differ somewhat from their standard values. If mating is at random with respect to age, and the other assumptions about the mating process which lead to equations (3.10) are fulfilled, we can write

$$k_{ij}^*(x, t) = k_{ij}^*(x)(1 + \xi_t)$$

where ξ_t is the same for each genotype, but in general changes with t. Using the assumption of weak selection, terms of order ξ_t^2 can be neglected, since differences in the male reproductive functions between different populations result from differences in the composition of the female population, and these are $O(\varepsilon)$.

Given these assumptions, the value of ξ_t at a time t close to some arbitrarily chosen fixed time t_0 is equal to $\xi_0 + O(\varepsilon^2)$, where ξ_0 is given by

$$(1 + \xi_0) \sum_x e^{-r_0 x} \sum_{ij} p_i(t_0) p_j(t_0) k_{ij}^*(x) = 1 \qquad (4.16)$$

(r_0 is defined on p. 156).

By the same type of argument that led to equation (4.11), it is possible to use equations (3.10) to obtain an approximate equation for the first difference in $\bar{p}_i = \frac{1}{2}(p_i + p_i^*)$. We define a new intrinsic rate r_{ij}' for A_iA_j males as the root of

$$(1 + \xi_0) \sum_x e^{-zx} k_{ij}^*(x) = 1$$

It turns out that $\Delta \bar{p}_i$ is determined by the $\bar{p}_j$ and the functions $\frac{1}{2}(r_{ij} + r'_{ij})$ in the same way that Δp_i in equation (4.11) is determined by the p_j and r_{ij}. Furthermore, if we write $T^*_{ij} = \sum x \exp[-r'_{ij}x]k^*_{ij}(x)$, it follows from the Taylor's expansion of the equation for r'_{ij} that

$$r'_{ij} = r^*_{ij} + (\xi_0 / T^*_{ij}) + O(\xi_0^2)$$

Neglecting terms in ε^2, r'_{ij} for any genotype is thus equal to r^*_{ij} plus a term common to each genotype. Hence, $\Delta \bar{p}_i$ is determined by the fitness functions $\bar{r}_{ij} = \frac{1}{2}(r_{ij} + r^*_{ij})$ for each genotype, to order ε^2. We have

$$\Delta \bar{p}_i = \bar{p}_i \left(\bar{r}_i - \sum_i \bar{p}_i \bar{r}_i \right) + O(\varepsilon^2) \tag{4.17}$$

where

$$\bar{r}_i = \sum_j \bar{p}_j \bar{r}_{ij}$$

The mean of the intrinsic rates of the males and females thus provides a measure of the fitness of a genotype. This is similar to the result for weak selection in the discrete-generation case, where the means of the male and female fitness values control gene frequency change (Wright, 1942; Nagylaki, 1979*b*).

The two-locus case. The two-locus case can be approached in a similar way to the one-locus model; using equations (3.17) we have for $t \geqq d$

$$B(t)p_i(t) = \sum_x B(t-x)p_i(t-x) \sum_j p_j(t-x)k_{ij}(x)$$

$$\pm c \sum_x B(t-x)D(t-x)k_{14}(x) \tag{4.18}$$

where the sign attached to c is positive for gametes 2 and 3, and negative for gametes 1 and 4.

When c is close to zero, the system can be treated as a multiple-allele system with four alleles, so that with weak selection, equations similar to (4.9), (4.11) and (4.15) must be obeyed. When linkage is loose in relation to the intensity of selection, so that $\varepsilon \ll c$, the following approach can be used, which is similar to that of Nagylaki (1976, 1977, pp. 169–173) for the discrete-generation case.

Writing $k_s(x)$ for the reproductive function of a standard genotype, as before, equation (4.18) becomes

$$B(t)p_i(t) = \sum_x B(t-x)p_i(t-x)k_s(x)$$

$$\pm c \sum_x B(t-x)D(t-x)k_s(x) + O(\varepsilon) \qquad (4.19)$$

Using equation (2.24) we can write

$$|D(t_0)| \leqq (1-c)|D(t_1)| + O(\varepsilon)$$

where t_0 and $t_1(t_1 < t_0)$ are two times separated by approximately one generation. Extending this back over a sequence of past times $t_2, t_3, \ldots t_l \ (t_{i+1} < t_i)$, as in equation (2.25), we obtain

$$|D(t_0)| \leqq (1-c)^l|D(t_i)| + [(1-c)^l + (1-c)^{l-1} + \ldots (1-c)]O(\varepsilon)$$

$$= (1-c)^l|D(t_i)| + \{(1-c)[1-(1-c)^l]/c\}O(\varepsilon)$$

Hence, if t_0 is taken sufficiently large, we have

$$|D(t_0)| \sim O(\varepsilon)$$

Substituting this into equation (4.19), we have for $t > t_0$

$$B(t)p_i(t) = \sum_x B(t-x)p_i(t-x)[k_s(x) + O(\varepsilon)]$$

Using the argument of Appendix 2, we conclude that $\Delta p_i \sim O(\varepsilon)$ and $\Delta^2 p_i \sim O(\varepsilon^2)$. Since $D \sim O(\varepsilon)$, ΔD for large t depends on the first differences of $O(\varepsilon)$ terms. By the methods of Appendix 2, these can be shown to be $O(\varepsilon^2)$, so that $\Delta D \sim O(\varepsilon^2)$. Using the same type of argument that led to equation (4.11), we can reduce equation (4.19) to the first-order form

$$\Delta p_i = p_i(r_i - \bar{r}) \pm (cD/T_s) + O(\varepsilon) \qquad (4.20)$$

where

$$T_s = \sum x \, e^{-r_s x} k_s(x)$$

This equation is similar to those for the discrete-generation case with selection weak in comparison with the recombination fraction (Nagylaki, 1976, 1977); the resemblance is perfect if both sides are multiplied by T_s, so that changes in gamete frequencies are measured in 'per generation' terms.

This completes the analysis of the case of weak selection with time-independent and density-independent demographic parameters. We have seen how it is possible to obtain approximate equations for this case which resemble closely the familiar equations for discrete-generation selection models, and that the intrinsic rate of increase of a genotype serves as an adequate measure of its fitness in this situation. It is even possible to handle the case of sex differences in demographic parameters, but the effects of non-random mating with respect to age have not been coped with. The significance of these results is discussed in the final section of this chapter. We turn next to the case when population size is limited by density-dependent factors.

4.2.2 *The density-dependent case*

The analysis of this situation is complicated by the fact that genetic changes due to selection must generally result in changes in population size, accompanied by changes in the density-dependent survival and fecundity components. It turns out, however, that one can obtain results which are very similar to those of the previous sections. The discussion here will be confined to the single-locus model obeying equations (3.14). Time-dependence of the $k_{ij}(x, t)$ is caused by their dependence on the numbers of individuals in the critical age-group at a given time or set of times, N_T, as described in sections 1.4.2 and 3.3.3.

Approximate equations for Δp_i. Consider a population at time t with gene frequencies $p_1(t), p_2(t), \ldots p_n(t)$ among the new zygotes. Assuming random mating with respect to age and genotype, it can be characterised demographically by its mean reproductive function $\sum_{ij} p_i(t)p_j(t)k_{ij}(x, N_T)$. We assume that a genetically homogeneous population with a reproductive function equal to this, and having the same functional relationship to N_T, would tend to a stationary size, with a number N_p of individuals in the critical age-group. It is assumed that such convergence to stationarity would take place for all reproductive functions which correspond to possible sets of gene frequencies, although the stationary population size N_p is, of course, a function of gene frequency, given by equation (3.41). It is also necessary to assume that, in a population with changing gene frequencies, the number of individuals in the critical age-group becomes equal to the equilibrium number for a standard genotype, plus terms of order ε, and the population growth-rate becomes $O(\varepsilon)$.† This can be used to show that Δp_i is asymptotically $O(\varepsilon)$, and that $\Delta^2 p_i$ is asymptotically $O(\varepsilon^2)$, on the lines of Appendix 2.

The number of individuals in the critical age-group at some time t sufficiently close to a fixed time t_0 can, from this, be approximated to $O(\varepsilon^2)$ by methods similar to those used in deriving equation (4.3). This number is given by

$$N(t) = \sum_{x \in S} B(t-x) \sum_{ij} p_i(t_0) p_j(t_0) [l_{ij}(x, N_T) + O(\varepsilon^2)] \qquad (4.21a)$$

and the total number of new zygotes satisfies

$$B(t) = \sum_x B(t-x) \sum_{ij} p_i(t_0) p_j(t_0) [k_{ij}(x, N_T) + O(\varepsilon^2)] \qquad (4.21b)$$

Equations (4.21) describe, to order ε^2, the dynamics of a genetically homogeneous population with reproductive function $\sum_{ij} p_i(t_0) p_j(t_0) k_{ij}(x, N_T)$. From the above assumptions about the properties of this case, it follows that

$$N(t) \sim N_0 + O(\varepsilon^2) \qquad (4.22a)$$

$$B(t-x) \sim B(t) + O(\varepsilon^2) \qquad (4.22b)$$

† This assumption is tantamount to the density-regulating mechanism being such that a genetically homogeneous population, exposed to continual, externally-imposed perturbations of order ε to its demographic parameters, will eventually settle down to a constant size, to $O(\varepsilon)$. This size corresponds to its equilibrium size in the absence of perturbations. For a treatment of a density-dependent, discrete-generation model which does not require such an assumption, see Nagylaki (1979a).

where N_0 is the value of N which satisfies $\sum_{ijx} p_i(t_0)p_j(t_0)k_{ij}(x, N) = 1$.

Carrying through the same type of manipulations as were used in the density-independent case, we obtain the gene-frequency equation analogous to equation (4.9)

$$T(N_p)\Delta p_i = p_i[w_i(N_p) - 1] + O(\varepsilon^2) \qquad (4.23)$$

We can make the same assumption as in Chapter 3: in a stationary population the net reproduction rate of each genotype is a strictly decreasing function of the number of individuals in the critical age-group. N_p is then given as the unique root of equation (3.41), and w_i and T in equation (4.23) are defined in the same way as in equations (4.8) and (4.9), but substituting $k_{ij}(x, N_p)$ for $e^{-r_p x}k_{ij}(x)$.

Equation (4.23) has an obvious resemblance to equation (4.9) for the density-independent case, and exactly the same properties hold for both, substituting the carrying-capacities N_{ij} and N_p of the density-dependent case for the r_{ij} and r_p of the other. There is thus at most one stable equilibrium with all alleles present, which corresponds to a maximum in N_p; with a system of two alleles, a stable equilibrium exists if and only if $N_{11}, N_{22} < N_{12}$, etc. Equation (4.23) can be approximated further on lines similar to those which yielded equation (4.11) for the density-independent case, by noting that

$$w_{ij}(N_p) = 1 + (N_p - N_{ij})\left(\frac{\partial w_{ij}}{\partial N}\right)_{N_{ij}} + O(\varepsilon^2)$$

Choosing an arbitrary genotype as standard, and writing T_s for the value of $\sum xk(x, N_T)$ for a stationary population with the demographic parameters of the standard genotype, and $\partial w_s/\partial N$ for the corresponding value of $(\partial w_{ij}/\partial N)_{N_{ij}}$, we obtain

$$\Delta p_i = -p_i \frac{(N_i - \bar{N})}{T_s}\left(\frac{\partial w_s}{\partial N}\right) + O(\varepsilon^2) \qquad (4.24)$$

where

$$N_i = \sum_j p_j N_{ij} \qquad \bar{N} = \sum_{ij} p_i p_j N_{ij}$$

Rate of change of population density. It is also possible to derive an equation for the rate of change of $N(t)$. Going through a similar procedure to that involved in obtaining equation (4.14), but using equation (4.24), we obtain the analogue of equation (4.15)

$$\Delta N = -2\left(\frac{\partial w_s}{\partial N}\right)\sum_i p_i \frac{(N_i - \bar{N})^2}{T_s} + O(\varepsilon^3) \qquad (4.25)$$

Equations similar to (4.23), (4.24) and (4.25) can be derived for the discrete-generation case (cf. Charlesworth, 1971; Roughgarden, 1976; León and Charlesworth, 1978; Kimura, 1978; Nagylaki, 1979a). They demonstrate that in a density-dependent population the progress of selection is controlled by the genotypic carrying-capacities, the N_{ij} (see equation (3.40)) and that selection tends to maximise the number of individuals in the critical age-group, subject to the assumption of weak selection. Numerical checks of these equations, based on computer calculations of population trajectories, show that, for moderate selection intensities, these equations provide accurate approximations for the rates of change of gene frequency and population density.

4.3   Local stability analyses

The results of the preceding section have shown that, with weak selection, it is possible to develop equations which give good approximations to the changes in gene frequencies and population size. It is obviously important to extend the theoretical analysis of the dynamics of selection to situations where there is arbitrarily strong selection, so that it is not legitimate to neglect second-order terms in the selection intensity. One way of tackling this problem is to carry out local stability analyses, by neglecting second-order terms in deviations of genotypic frequencies from their equilibrium values, in the neighbourhoods of equilibria. The present section describes this approach; the account is based mainly on the papers of Charlesworth (1973, 1976). Pollak and Kempthorne (1971) have also studied this problem. In addition to local analyses based on the deterministic equations, the stochastic behaviour of a mutant gene in a large population is examined by branching process methods. Only the case of a single autosomal locus with two alleles has been studied.

4.3.1 *Gene frequencies near 0 or 1 with density independence*

A rare, non-recessive allele. This case may be exemplified by considering the fate of allele A_2 when introduced at a low frequency into a population fixed for A_1. Terms involving the square of the frequency of A_2 can be neglected in the initial stages after its introduction, while it remains rare. In principle, this enables us to deal with the case when mating is non-random with respect to age, since by inspection of equations (3.21), we can neglect terms arising from the variance in gene frequency across age-classes, and the second-order terms in the difference between the male and female gene frequencies. This enables us to use equations (3.10) even with non-random mating with respect to age. We have for $t \geqq d$

$$B(t) = \sum_x B(t-x) k_{11}(x) + O(p_2) \qquad (4.26)$$

so that asymptotically (cf. Appendix 2)

$$B(t-x) \sim B(t) e^{-r_{11}x} + O(p_2) \qquad (4.27)$$

Substituting this into equations (3.10*a*) and (3.10*b*), for the two-allele case with time-independent reproductive functions, gives (cf. p. 162)

$$\bar{p}_2(t) = \tfrac{1}{2} \sum_x e^{-r_{11}x} \bar{p}_2(t-x) [k_{12}(x) + k_{12}^*(x)] + O(\bar{p}_2^2) \qquad (4.28)$$

Note that the $k_{12}^*(x)$ here are the values derived using the age-specific fecundities of A_1A_2 males when mated to females drawn from a wholly A_1A_1 population in stable age-distribution. To determine their values in a given case, one would need to specify in detail the rules of mating with respect to age and genotype, or to have empirical data on the fecundities of rare A_1A_2 males in a largely A_1A_1 population.

The logarithmic rate of increase in frequency of A_2 is therefore given asymptotically, to order $O(\bar{p}_2)$, by the real root of

$$\tfrac{1}{2} \sum_x e^{-(z+r_{11})x} [k_{12}(x) + k_{12}^*(x)] = 1 \qquad (4.29)$$

The utility of this result depends on the approach to the asymptotic values for $B(t)$ and $\bar{p}_2(t)$ being sufficiently fast that the latter is still small enough for terms of order $\bar{p}_2^2$ to be negligible after convergence has taken place. As discussed in section 1.3.2, convergence to the asymptotic solutions of equations of the same form as equations (4.26) and (4.28) is normally a matter of a few generations. Provided $\bar{p}_2$ is initially fairly small, there should be no problem. Numerical examples indicate that in practice the result can be used for initial $\bar{p}_2$ values of up to 0.01 or so, even with strong selection.

When mating is at random with respect to age and genotype, and when the assumptions of equations (3.13) are satisfied (no mortality differences between males and females of the same age and genotype, and proportionality of male and female fecundities), it is possible to show that $m_{12}^*(x) = m_{12}(x)$ plus terms of order p_2. $k_{12}(x)$ can therefore be substituted for $k_{12}^*(x)$ in equation (4.29), so that for large t

$$\Delta \ln p_2 = (r_{12} - r_{11}) + O(p_2) \tag{4.30}$$

Obviously, the same analysis can be carried out for the other endpoint in gene frequency, $p_1 = 0$. The rate of spread of A_1 when introduced at a low frequency into an $A_2 A_2$ population is governed by equations similar to (4.29) and (4.30), except that r_{22} is substituted for r_{11}, p_1 for p_2, and the values of the $k_{12}^*(x)$ are those for mating $A_1 A_2$ males to females from a stable $A_2 A_2$ population.

A point of interest which emerges from this analysis is that, for the case of random mating with respect to age and no demographic differences between males and females (equation (4.30)), both endpoints of gene frequency are unstable if there is heterozygote superiority in the intrinsic rate of increase. In other words, each allele tends to increase in frequency when rare, so that there is a *protected polymorphism* in the sense of Prout (1968), when $r_{12} > r_{11}$, r_{22}. If the reverse inequality holds, both alleles tend to be eliminated when rare. This corresponds to the results of the analysis for weak selection, and to the standard results for discrete generations if the genotypic fitnesses W_{ij} are substituted for the r_{ij}. It suggests that, with strong selection, there may be convergence to the equilibrium given by equations (3.38) and (3.39) when there is heterozygote advantage in the r_{ij}. If $r_{11} < r_{12} < r_{22}$, the endpoint $p_1 = 1$ is unstable, and $p_2 = 1$ is stable, which suggests that A_2 spreads towards fixation. The extent to which these conjectures are true is discussed in sections 4.3.5 and 4.4.

A rare, recessive allele. If either $r_{11} \cong r_{12}$ or $r_{22} \cong r_{12}$, the above approach breaks down when A_2 or A_1, respectively, is rare, since the initial rate of spread of the recessive allele is of the same order as the square of its frequency. A different method of approximation must therefore be used when a rare allele is recessive in its effect on the intrinsic rate of increase. We examine here equation (4.5) for a two-allele system, with A_2 introduced at low frequency into an A_1A_1 population and with $r_{12} = r_{11}$. Using the result that Δp_2 is $O(p_2^2)$, we can employ the same type of approach as for equations (4.2), (4.3) and (4.4) to obtain

$$B(t) = \sum_x B(t-x)\{[1 - 2p_2(t_0)]k_{11}(x)$$

$$+ 2p_2(t_0)k_{12}(x)\} + O(p_2^2) \qquad (4.31)$$

where t_0 is a fixed time not too far from t. (Note that this does not assume anything about the strength of selection.) Since $r_{12} = r_{11}$, we thus have

$$B(t-x) \sim B(t)\, e^{-r_{11}x} + O(p_2^2)$$

Substituting this into equation (4.5) for allele A_2, we obtain

$$p_2(t) \sim \sum_x e^{-r_{11}x} p_2(t-x)[p_1(t-x)k_{12}(x) + p_2(t-x)k_{22}(x)] + O(p_2^3) \qquad (4.32)$$

Taking first differences on both sides, we have

$$\Delta p_2 \sim \sum_x e^{-r_{11}x} \Delta p_2(t-x)k_{12}(x) + O(p_2^3)$$

so that Δp_2 approaches a constant value asymptotically, to order p_2^3 (cf. Appendix 2). The second-order and higher-order differences in p_2 can therefore be neglected, so that equations (4.32) can be expanded on similar lines to equations (4.6) and (4.7), to yield the final result

$$\Delta p_2 \sim \frac{p_2^2[w_{22}(r_{11}) - 1]}{T_{12}(r_{11})} + O(p_2^3) \qquad (4.33)$$

If selection is weak, we can use the argument that led to equation (4.11); neglecting terms $O(\varepsilon^2)$, we obtain

$$\Delta p_2 \sim p_2^2(r_{22} - r_{11}) \qquad (4.34)$$

If selection is strong, however, equations (4.33) and (4.34) yield different values for the change in gene frequency. Comparisons with the values of Δp_2 obtained by exact calculations of population trajectories show that equation (4.33) is accurate with strong selection, provided that p_2 is small (<0.05 or so), whereas equation (4.34) breaks down with strong selection, although A_2 can spread only when $r_{22} > r_{11}$. We therefore have the somewhat surprising result that when selection is strong, the *rate* of spread of a rare recessive gene cannot be calculated purely from its effect on the intrinsic rate of increase, although the *direction* is determined by this quantity. This contrasts with the results for the case of a gene with a significant effect on the intrinsic rate of increase of the heterozygote.

4.3.2 *Gene frequencies near 0 or 1 with density dependence*

The results of the previous two sections can easily be extended to the density-dependent case. The introduction of a rare allele will, as before, perturb the demographic parameters of the population by an amount of the order of the gene frequency (or square of the gene frequency in the case of a recessive allele). If the density-regulating mechanism is such that a genetically homogeneous population tends to a stationary size, as is assumed throughout our treatment of density-dependence, the analogues of equation (4.27) for this density-dependent case with a rare, non-recessive gene are

$$B(t-x) \sim B(t) + O(p_2) \qquad (4.35a)$$

$$N(t) \sim N_{11} + O(p_2) \qquad (4.35b)$$

Substituting from these equations into the gene frequency equations (3.14), we find that the asymptotic logarithmic rate of increase in the frequency of A_2 is, to order p_2, given by the real root of the equation

$$\sum_x e^{-zx} k_{12}(x, N_{11}) = 1 \qquad (4.36)$$

It is easy to see that for this root to be positive we must have $N_{12} > N_{11}$ in the case when the $w_{ij}(N)$ as defined in equation (3.40) are strictly decreasing functions of N. Hence, a rare gene spreads if the carrying-capacity of the heterozygote exceeds that of the initial population. Applying this to the other endpoint gives the result that there is a protected polymorphism if $N_{12} > N_{11}, N_{22}$. This is similar to the results derived in section 4.2.2 for the case of weak selection, but is obviously more restricted, since nothing has been proved about convergence to the equilibrium.

A similar type of argument can be applied to the problem of the introduction of a rare recessive gene. The analogue of equation (4.33) is

$$\Delta p_2 \sim \frac{p_2^2[w_{22}(N_{11}) - 1]}{T_{12}(N_{11})} + O(p_2^3) \qquad (4.37)$$

A recessive gene, therefore, spreads if and only if the carrying-capacity of the homozygote exceeds that of the initial population, in the case when the $w_{ij}(N)$ are decreasing functions of N.

4.3.3 *Gene frequencies near 0 or 1 with temporally varying environments*

The results derived up to now have assumed that the demographic parameters of each genotype are independent of time, or that they depend on time solely via changes in population density. There has recently been much interest in the population genetics of selection in temporally varying environments (e.g. Haldane and Jayakar, 1963; Gillespie, 1973*a, b*, 1978; Karlin and Lieberman, 1974), so that it is natural to enquire how far the results can be extended to age-structured populations. The discussion here is confined to the question of the conditions for maintenance of a protected polymorphism in a temporally varying environment. The nature of the method of analysis means, as will be seen, that the population is density independent. As previously, the gene frequency equations (3.14) are used.

A rare, non-recessive allele. For A_2 introduced into an $A_1 A_1$ population at a low frequency, we have

$$B(t)p_2(t) = \sum_x B(t-x)p_2(t-x)k_{12}(x, t) + O(p_2^2) \qquad (4.38a)$$

$$B(t) = \sum_x B(t-x)k_{11}(x, t) + O(p_2) \qquad (4.38b)$$

The process represented by these equations can be expressed in terms of the equivalent Leslie matrix formulations, using the matrices for A_1A_1 and A_1A_2, which we can write as $L_{11}(t)$ and $L_{12}(t)$ for time t (cf. equation (1.6)). Since we have assumed density independence we can use the weak ergodicity theorem of Lopez (1961) (section 1.4.1) to deduce that for sufficiently large t

$$B(t)p_2(t) = \lambda_{12}(t-1)B(t-1)p_2(t-1) + O(p_2^2) \qquad (4.39a)$$

$$B(t) = \lambda_{11}(t-1)B(t-1) + O(p_2) \qquad (4.39b)$$

where $\lambda_{11}(t-1)$ is the geometric growth rate of a homozygous A_1A_1 population from time $t-1$ to t, and $\lambda_{12}(t-1)$ is the rate for a population with the demographic parameters of A_1A_2.

Combining these two equations, we obtain the analogue of equation (4.30)

$$\Delta \ln p_2(t) = [r_{12}(t) - r_{11}(t)] + O(p_2) \qquad (4.40)$$

where $r_{ij}(t) = \ln \lambda_{ij}(t)$, and measures the increase in $\ln B$ between t and $t+1$, for a population with the demographic parameters of A_iA_j.

From the weak ergodicity theorem, the $r_{ij}(t)$ and $\lambda_{ij}(t)$ are independent of the initial conditions, for sufficiently large t, and are determined only by the environmental process which generates the time dependence of the demographic parameters. The difficulty with this result is that it is not generally possible to obtain an analytic expression for the values of the r_{ij} and λ_{ij} as functions of time. The following useful conclusions can, however, be drawn from equation (4.40).

In the first place, equation (4.40) tells us that the mean change in gene frequency between generations t and $t+n$ for large t is given by

$$\frac{1}{n}[\ln p_2(t+n) - \ln p_2(t)] = \frac{1}{n}\left\{\sum_{j=0}^{n-1} [r_{12}(t+j) - r_{11}(t+j)]\right\} + O(p_2)$$

$$(4.41)$$

so that, to order p_2^2, there will be an average increase in the frequency of A_2 if an A_1A_2 population has a higher arithmetic mean value of $r_{ij}(t)$ than an A_1A_1 one. This is analogous to the result of Haldane and Jayakar (1963) for the corresponding discrete-generation case, with the $r_{ij}(t)$ in the age-structured case playing the same role as the logarithms of the genotypic fitnesses with discrete generations. It implies that, in a deterministically varying environment, a rare gene will increase in frequency asymptotically if a population with the demographic parameters of the heterozygote has a higher mean rate of increase than the population homozygous for the original allele.

A similar conclusion can be drawn for the case of a stochastically varying environment by using the theorem of Cohen (1976, 1977a, b) that the probability distribution of age-structure, and hence rate of increase of a population at a given time t, converges to a stationary distribution when exposed to Markovian temporal variation in its demographic parameters (cf. section 1.4.1). This implies that the $r_{ij}(t)$ in equation (4.41) can be regarded as sampled from such a stationary distribution, provided that the assumptions of the theorem apply to the matrices $L_{11}(t)$ and $L_{12}(t)$. This enables us to apply the arguments used for the discrete-generation case by Gillespie (1973a) and Karlin and Lieberman (1974). We can conclude that there is a probability close to one that p_2 is bounded away from zero (i.e. the endpoint $p_1 = 1$ is *stochastically unstable*) if the expectation of $r_{12}(t) - r_{11}(t)$, over the stationary distribution generated by the environmental process, is positive.

These results can obviously be extended to the other endpoint $p_1 = 0$, provided that A_1 is not recessive to A_2; a protected polymorphism is maintained, therefore, if the arithmetic mean of the $r_{ij}(t)$ for a population with the demographic characteristics of the heterozygote exceeds the means for the homozygotes.

A rare, recessive allele.   When one allele, say A_2, is recessive to the other, the above approach breaks down for the same reasons as in the time-independent case. In the discrete-generation case, Haldane and Jayakar (1963) showed that a protected polymorphism can be maintained at a locus where one allele is completely recessive to the other, as a result of temporal fluctuations in fitnesses. This requires the arithmetic mean of the fitness of the homozygote for the recessive allele (A_2) to exceed that of A_1A_1 and A_1A_2, whereas its geometric mean must be less than the geometric mean for A_1A_1 and A_1A_2. Unfortunately, it does not seem possible to provide a general analysis similar to this for the age-structured case. The following special case proves, however, that the Haldane–Jayakar mechanism can work in an age-structured context.

Consider the situation in which there are environmental states which alternate at intervals which are very long compared with the life-span of an individual. For most of the time between the environmental changes, a population in such a situation can be described accurately by the gene frequency equations with the demographic parameters which are appropriate for the given environmental state in which it finds itself, since the transitional states created by a change in environment occupy a negligible part of the population's history. Hence, for a population in environmental state i, the rate of gene frequency change for A_2 when rare is approximated accurately for most of the time by equation (4.33) with the demographic parameters for environment i

$$\Delta p_2 \sim \frac{p_2^2[w_{22}^{(i)}(r_{11}^{(i)})-1]}{T_{12}^{(i)}(r_{11}^{(i)})}+O(p_2^3) \tag{4.42}$$

where $r_{11}^{(i)}$ is the intrinsic rate of increase of an A_1A_1 population maintained in environment i indefinitely, etc. By the same type of argument as employed by Haldane and Jayakar (1963), p_2 tends to increase in frequency in the long term if the arithmetic mean of

$$\frac{w_{22}^{(i)}(r_{11}^{i})-1}{T_{12}^{(i)}(r_{11}^{(i)})}$$

taken over all the environmental states, exceeds zero. The same type of argument concerning the effects of the long environmental period can be applied to the other endpoint; combined with the results of section 4.3.1, we find that A_1 increases when rare if the arithmetic mean of $r_{12}^{(i)} - r_{22}^{(i)}$ is positive. The existence of a protected polymorphism requires both these conditions to be satisfied simultaneously.

If selection is weak, the condition for the increase of A_2 reduces to requiring the arithmetic mean of $r_{12}^{(i)} - r_{22}^{(i)}$ to be negative (cf. equation (4.34)). This is obviously impossible if the condition for the increase of A_1 is satisfied. We can conclude that a protected polymorphism cannot exist when selection is weak. (This is also true for the discrete-generation case, since the arithmetic and geometric means of the fitnesses approach each other for weak selection.) When selection is strong, however, it is possible for the condition for the spread of A_2 to hold when the mean of $r_{12}^{(i)} - r_{22}^{(i)}$ is positive, so that a protected polymorphism can exist under suitable forms of environmental fluctuation. An extreme example of how this is possible is as follows. Suppose there are just two environmental states, in one of which A_2A_2 is near-lethal, so that r_{22} for this state is very large and negative. Provided that r_{22} for the other environment is bounded, the mean of $r_{12}^{(i)} - r_{22}^{(i)}$ will be positive if A_1A_1 and A_1A_2 have reasonable viability in both environments. If, however, A_2A_2 has a sufficiently high survival in the second environment compared with A_1A_1 and A_1A_2, the criterion for the increase of A_2 when rare can be satisfied, despite the near-lethality of A_2A_2 in the first environment.

Computer calculations of population trajectories have shown that this approximate method predicts the conditions for maintenance of polymorphism when the environmental period is of the order of several generations, and that there is no more difficulty in generating numerical examples which satisfy the conditions than in the discrete generation case.

The conditions for the spread of a rare gene, and hence for protected polymorphism, are less clear for the case when the environmental period is of the order of or smaller than the life span, or when there is stochastic variation in the environment. In such cases, the response of the population to changing demographic parameters means that the rare gene can never settle down to a constant rate of increase in frequency. All that can be done with present theory is to calculate the asymptotic growth rates, the $r_{ij}(t)$, for specific numerical examples. Templeton and Levin (1979) discuss the analysis of selection with fluctuating environments for the special case of plants with seed dormancy (cf. pp. 8–10 and 46–7).

4.3.4 *The probability of survival for a non-recessive mutant gene*
The above results are based on the assumption that the spread of a rare gene in an infinite population can be satisfactorily described from a purely deterministic viewpoint. In actual fact, a rare gene is represented by so few copies in the initial generations after its occurrence by mutation that even in an infinite population it

is subject to stochastic fluctuations and a significant probability of loss. This process was first studied by Fisher (1922, 1930) and Haldane (1927*b*) for the case of a discrete-generation population. In this section, we shall study the problem of the probability of survival of a mutant gene in an infinite population with time-independent demographic parameters. The account is based on the treatment of Charlesworth and Williamson (1975). A somewhat more general treatment is given by Pollak (1976). The results will be used in Chapter 5 (pp. 211–4).

We again use the model expressed in equation (4.5), which assumes random mating with respect to genotype and age, equality of age-specific survivorships of males and females of the same genotype, and proportionality of male and female age-specific fecundities. If A_2 is introduced by mutation into an A_1A_1 population, it will initially be overwhelmingly represented in A_1A_2 heterozygotes. Furthermore, if the initial A_1A_1 population is in stable age-distribution, or is held at a stationary equilibrium size by density-dependent factors, we saw on p. 170 that $m_{12}^*(x) = m_{12}(x)$ plus terms of order p_2. It is therefore reasonable to assume that, neglecting terms $O(p_2)$, the distributions of numbers of offspring for A_1A_2 males and females of a given age are identical, provided that they are completely characterised by their mean (e.g. when they are Poisson). This assumption is made from now on, and enables us to disregard the sex of individuals. A final assumption is that the primary sex-ratio is equal to one-half. This implies that the mean number of A_1A_2 offspring, of either sex, of an A_1A_2 heterozygote is equal to the mean number of its offspring that are the same sex as itself, which enables us to simplify the notation.

A new mutant gene A_2 will first manifest itself as an A_1A_2 zygote. The problem therefore reduces to that of finding the probability of survival of a line of A_1A_2 individuals descended from a single A_1A_2 zygote. Let this probability be U. We can arbitrarily classify the set of progeny produced by a given zygote during the whole of its lifetime as belonging to the same 'generation', although many of them will in fact have been produced in different time-intervals. Standard branching-process theory (Feller, 1968, Chapter 12) can thus be applied to successive 'generations' descended from the initial zygote carrying A_2. This yields the probability of loss of A_2 as the smallest value of s, in the interval $0 \leqq s \leqq 1$, which satisfies the equation

$$h_{12}(s) = s \qquad (4.43)$$

where $h_{12}(s)$ is the *probability generating function* (p.g.f.) of the distribution of the number of A_1A_2 offspring produced by an A_1A_2 individual over the whole of its life-span (cf. Appendix 1). It follows from standard theory that the mutant gene has a non-zero probability of survival if and only if the mean of this offspring distribution exceeds one, i.e. the net reproduction rate, $R_{12} = \sum k_{12}(x)$, exceeds one. This is also the condition for $r_{12} > 0$.

The functional form of $h_{12}(s)$ can be obtained as follows. Let the p.g.f. for the number of A_1A_2 offspring produced by an A_1A_2 parent aged x be $h_{12}(x, s)$. Assuming that, for a parent who has survived to a given age, the offspring distributions for different earlier ages are independent, the p.g.f. for the number of A_1A_2 offspring produced between ages b and x by an individual now aged $x (x \geqq b)$ is

$$H_{12}(x, s) = \prod_{y=b}^{x} h_{12}(y, s) \qquad (4.44)$$

The probability that an A_1A_2 individual survives to age x and dies between x and $x+1$ is $l_{12}(x) - l_{12}(x+1)$. By partitioning the life-history of an individual into mutually exclusive events corresponding to these probabilities for all different x values, it is easy to see that

$$h_{12}(s) = 1 - l_{12}(b) + \sum_{x=b}^{d} [l_{12}(x) - l_{12}(x+1)] H_{12}(x, s) \quad (4.45)$$

In the important case when the offspring distribution for each age-class is Poisson, so that $h_{12}(x, s) = \exp[(s-1)m_{12}(x)]$, the survival probability U, from equation (4.43), satisfies the equation

$$U + \sum_{x} [l_{12}(x) - l_{12}(x+1)] \exp[-\tilde{M}_{12}(x)U] = l_{12}(b) \quad (4.46)$$

where $\tilde{M}_{12}(x) = \sum_{y=b}^{x} m_{12}(y)$.

As shown by Charlesworth and Williamson (1975), the survival probability given by equation (4.46) is always less than that for a discrete generation model with a Poisson offspring distribution and the same mean. These authors give an example for a human population where the reduction is as high as 40 % of the discrete generation value.

For U close to zero and R_{12} close to unity, equation (4.43) can be approximated in the standard way (e.g. Ewens, 1969, p. 80) to give

$$U \cong 2(R_{12}-1)/V_{12} \qquad (4.47)$$

where V_{12} is the variance of the lifetime offspring distribution. In the case of a Poisson distribution for each age-class,

$$V_{12} \cong \sum [l_{12}(x) - l_{12}(x+1)]\tilde{M}_{12}^2(x)$$

when $R_{12} \cong 1$.

Equation (4.47) may be thought of as giving a useful approximation to the survival probability of a mutant gene with a small effect on demographic parameters in the heterozygous state, when introduced into a stationary or near-stationary population. Using the approximation for the intrinsic rate of increase (equation (1.46)), equation (4.47) can be rewritten as

$$U \cong 2r_{12}T_{12}/V_{12} \qquad (4.48)$$

where $T_{12} = \sum xk_{12}(x)$. It will be seen from these equations that small changes in the demographic parameters which affect R_{12} and r_{12} alone have much bigger effects on U than changes which affect V_{12} or T_{12}, provided that U is small. This implies that the probability of survival of a mutant gene in a near-stationary population is largely controlled by its effect on R_{12} or, equivalently, r_{12}.

4.3.5 *Local stability of a polymorphic equilibrium*

The purpose of this section is to extend the method of local stability analysis to the polymorphic equilibrium which exists in the two-allele case when there is heterozygote superiority or inferiority in the intrinsic rates of increase, for the time-independent case (Chapter 3, pp. 136–8). We saw on p. 158 of this chapter that heterozygote superiority in the r_{ij} corresponds to global stability of the equilibrium in the case of weak selection, and heterozygote inferiority to instability. We show here that similar results hold for local stability with arbitrary selection intensities.

Derivation of a stability criterion. The equilibrium gene frequencies $\hat{p}_1$ and $\hat{p}_2 = 1 - \hat{p}_1$ are given by equation (3.38), and the equilibrium population growth-rate by equation (3.39). We can linearise equations (3.14), for the two-allele case, by neglecting terms in $(p_1 - \hat{p}_1)^2$. This will give an adequate approximation for gene frequencies close to their equilibrium values. Writing $B_1(t) = B(t)[p_1(t) - \hat{p}_1]$, we obtain

$$B(t)\hat{p}_1 + B_1(t) = g_1(t) + \hat{p}_1 \sum_{x=1}^{t} B(t-x)k_1(x)$$

$$+ \sum_{x=1}^{t} B_1(t-x)K_1(x) \qquad (4.49a)$$

$$B(t)\hat{p}_2 - B_1(t) = g_2(t) + \hat{p}_2 \sum_{x=1}^{t} B(t-x)k_2(x)$$

$$- \sum_{x=1}^{t} B_1(t-x)K_2(x) \qquad (4.49b)$$

where

$$k_1(x) = \hat{p}_1 k_{11}(x) + \hat{p}_2 k_{12}(x), \quad K_1(x) = 2\hat{p}_1 k_{11}(x) + (1 - 2\hat{p}_1)k_{12}(x)$$

$$k_2(x) = \hat{p}_1 k_{12}(x) + \hat{p}_2 k_{22}(x), \quad K_2(x) = 2\hat{p}_2 k_{22}(x) - (1 - 2\hat{p}_1)k_{12}(x)$$

and where $g_1(t)$ and $g_2(t)$ are zero for $t \geq d$.

Taking generating functions on both sides of these equations (see Appendix 1 for the properties of generating functions which are used here), we obtain

$$\hat{p}_1 \bar{B}(1 - \bar{k}_1) + \bar{B}_1(1 - \bar{K}_1) = \bar{g}_1 \qquad (4.50a)$$

$$\hat{p}_2 \bar{B}(1 - \bar{k}_2) - \bar{B}_1(1 - \bar{K}_2) = \bar{g}_2 \qquad (4.50b)$$

These equations are easily solved for $\bar{B}$ and $\bar{B}_1$:

$$\bar{B} = [\bar{g}_1(1 - \bar{K}_2) + \bar{g}_2(1 - \bar{K}_1)]/\Delta \qquad (4.51a)$$

$$\bar{B}_1 = [\hat{p}_2 \bar{g}_1(1 - \bar{k}_2) - \hat{p}_1 \bar{g}_2(1 - \bar{k}_1)]/\Delta \qquad (4.51b)$$

where

$$\Delta = \hat{p}_1(1 - \bar{k}_1)(1 - \bar{K}_2) + \hat{p}_2(1 - \bar{k}_2)(1 - \bar{K}_1) \qquad (4.52)$$

Using the results described in Appendix 1, it follows that for sufficiently large t we have

$$B(t) = C_1 s_1^{-(t+1)} \qquad (4.53a)$$

where s_1 is the zero of Δ in equation (4.52) with smallest modulus, excluding any common zeros of $1 - \bar{K}_1$ and $1 - \bar{K}_2$.

Similarly, for large t we have

$$B_1(t) = C_2 s_2^{-(t+1)} \qquad (4.53b)$$

where s_2 is the zero of Δ with smallest modulus, excluding any common zeros of $1 - \bar{k}_1$ and $1 - \bar{k}_2$. Note that it is possible for s_1 and s_2 to be equal. Equations (4.53) assume that s_1 and s_2 are simple zeros. The modifications to these equations when s_1 or s_2 is a repeated zero of Δ are relatively straightforward (see Appendix 1), and do not materially affect the results described below.

As with the renewal treatment of the case of a genetically homogeneous population (equation (1.45)), it is more convenient in practice to work with the natural logarithms of s_1^{-1} and s_2^{-1}, which we denote by z_1 and z_2. Equations (4.53) thus become

$$B(t) = C_1 e^{z_1(t+1)} \qquad (4.54a)$$

$$B_1(t) = C_2 e^{z_2(t+1)} \qquad (4.54b)$$

If z_1 is a complex number, then equation (4.54a) must be replaced by

$$B(t) = C_1 e^{z_1(t+1)} + C_1^* e^{z_1^*(t+1)} \qquad (4.55)$$

where z_1^* and C_1^* are complex conjugates of z_1 and C_1, respectively. A corresponding equation replaces (4.54b) if z_2 is complex.

If we can show that the real part of z_1 is equal to $\hat{r}$, and that the real part of z_2 is less than $\hat{r}$, the equilibrium can be said to be locally stable, since $B(t)$ must return to its equilibrium rate of growth, and p_1 to $\hat{p}_1$. Conversely, if the real parts of z_1 and/or z_2 exceed $\hat{r}$, the equilibrium must be locally unstable.

Unfortunately, it is difficult to give an analysis of the zeros of Δ analogous to that of equation (1.32), owing to the large number of terms in equation (4.52). An indirect approach can, however, be used to establish that, when there is heterozygote superiority in the intrinsic rates of increase, the equilibrium is locally stable.

Stability when $r_{12} > r_{11}, r_{22}$. We first note that $\hat{r}$ is, from equations (3.37), the root of both $1 - \bar{k}_1 = 0$ and $1 - \bar{k}_2 = 0$ with largest real part (from now on, the generating functions are treated as functions of $z = \ln(s^{-1})$ rather than s). Assume arbitrarily that $\hat{p}_2 > \hat{p}_1$. Then $K_1(x)$ is non-negative, from equations (4.49), and the zero of $1 - \bar{K}_1$ with largest real part, denoted by z', is real by the argument of pp. 25–7. The formula for $K_1(x)$ also implies that $r_{11} < z' < r_{12}$. Furthermore, we can write

$$K_1(x) = k_1(x) + \hat{p}_1[k_{11}(x) - k_{12}(x)]$$

This implies $z' < \hat{r}$ since it is easily seen from this formula that $\sum e^{-\hat{r}x} K_1(x) < 1$.

If the real parts of z_1 and z_2 are greater than $\hat{r}$, we must have $z_1 = z_2$, since only the common zeros of $1 - \bar{K}_1$ and $1 - \bar{K}_2$ or $1 - \bar{k}_1$ and $1 - \bar{k}_2$ are *not* simultaneous roots for $\bar{B}$ and $\bar{B}_1$, and we have just established that only $1 - \bar{K}_2$ can have a zero with real part greater than $\hat{r}$. Hence, *either* the real parts of z_1 and z_2 are smaller than or equal to $\hat{r}$, *or* $z_1 = z_2$.

It can be established as follows that there can be no *real* zero of Δ which is greater than $\hat{r}$. Rearranging equation (4.52) gives

$$\Delta = \hat{p}_1(1 - \bar{k}_{11})(1 - \bar{k}_2) + \hat{p}_2(1 - \bar{k}_{22})(1 - \bar{k}_1) \qquad (4.56)$$

It is easy to see that, for all real $z > \hat{r}$, $\Delta > 0$, so that Δ cannot have a real zero greater than $\hat{r}$. The only way in which the equilibrium can be unstable in this case is if there is a complex zero of Δ, $z_1 = z_2$, with a real part greater than $\hat{r}$. The following argument shows that this situation cannot occur.

Substituting from equation (4.55) into equation (4.49a), we obtain

$$\hat{p}_1 C_1 e^{z_1(t+1)}(1 - \bar{k}_1) + \hat{p}_1 C_1^* e^{z_1^*(t+1)}(1 - \bar{k}_1^*) + C_2 e^{z_1(t+1)}(1 - \bar{K}_1)$$
$$+ C_2^* e^{z_1^*(t+1)}(1 - \bar{K}_1^*) = 0$$

where

$$\bar{k}_1^* = \sum e^{-z_1^* x} k_1(x), \qquad \bar{K}_1^* = \sum e^{-z_1^* x} K_1(x)$$

If we write $z_1 = \alpha + i\beta$, this gives

$$e^{2i\beta(t+1)}[\hat{p}_1 C_1(1 - \bar{k}_1) + C_2(1 - \bar{K}_1)] + \hat{p}_1 C_1^*(1 - \bar{k}_1^*) + C_2^*(1 - \bar{K}_1^*)$$
$$= 0 \qquad (4.57)$$

This equation must be satisfied for arbitrary t, which is impossible unless z_1 is a zero of both $1 - \bar{k}_1$ and $1 - \bar{K}_1$; this possibility has been excluded already. Hence, there cannot be a complex zero of equation (4.52) with its real part greater than $\hat{r}$.

This completes the results which are necessary to establish that the polymorphic equilibrium is locally stable if $r_{12} > r_{11}, r_{22}$. Since there is no complex zero of Δ with real part greater than $\hat{r}$, it follows that z_1, the largest root for $B(t)$, must equal $\hat{r}$ (see equation (4.52)), and that z_2, the largest root for $B_1(t)$, must have its real part less than $\hat{r}$. From the definition of $B_1(t)$, the deviation of the gene frequency from its equilibrium value, $p_1 - \hat{p}_1$, must asymptotically be proportional to $\exp[(z_2 - \hat{r})t]$, and hence tend to zero. Convergence of the gene frequency to equilibrium takes place via a series of damped oscillations. The population growth-rate similarly converges to its equilibrium value $\hat{r}$. Computer calculations of population trajectories show this pattern clearly (see, for example, Figure 2 of Anderson and King, 1970). The rate of convergence is determined by the difference between the real part of z_2 and $\hat{r}$. In general, z_2 can be determined only by numerical evaluation of the zeros of Δ. The convergence to equilibrium via damped oscillations in this case contrasts with the result for discrete generations with heterozygote superiority in fitness, where convergence to the equilibrium is monotonic (cf. Nagylaki, 1977, pp. 56–8).

Instability when $r_{12} < r_{11}, r_{22}$. In this case, it is easy to see from equation (4.56) that there is a real zero of Δ which exceeds $\hat{r}$. An argument similar to the one used above shows that there cannot be a complex zero with real part greater than this. The population in the neighbourhood of the equilibrium thus asymptotically moves away from the equilibrium point at a constant rate.

4.4 The asymptotic results of selection

Some theoretical results concerning the ultimate states reached by populations under arbitrarily strong selection intensities have been obtained by Norton (1928), Pollak and Kempthorne (1971) and Charlesworth (1974a). These results all pertain to the two-allele versions of equation (4.5), with time-independent and density-independent demographic parameters, or to the corresponding continuous-time equations. The most complete analysis was given by Norton for the continuous-time model. His analytical methods are too complex to describe here, but his conclusions are summarised in Table 4.1, along with the corresponding discrete-generation results. They agree with the results derived in section

Table 4.1. *The outcomes of various types of selective regimes with a diallelic locus*

Discrete-generation populations		Age-structured populations[a]	
Selection regime	Outcome of selection	Selection regime	Outcome of selection
$w_{11} = w_{12} = w_{22}$	p_1 constant (neutral equilibrium)	$r_{11} = r_{12} = r_{22}$	p_1 constant, or neutral oscillations
$w_{11} < w_{12} \leqq w_{22}$	$p_1 \to 0$	$r_{11} < r_{12} \leqq r_{22}$	$p_1 \to 0$
$w_{11} \leqq w_{12} < w_{22}$	$p_1 \to 0$	$r_{11} \leqq r_{12} < r_{22}$	$p_1 \to 0$
$w_{12} > w_{11}, w_{22}$	$p_1 \to \hat{p}_1$ $(0 < \hat{p}_1 < 1)$	$r_{12} > r_{11}, r_{22}$	$p_1 \to \hat{p}_1$, or sustained oscillations about $\hat{p}_1$
$w_{12} < w_{11}, w_{22}$	$p_1 \to 1$, or $p_1 \to 0$	$r_{12} > r_{11}, r_{22}$	$p_1 \to 1$ or $p_1 \to 0$

[a] These results are due to Norton (1928), and were derived for a continuous-time model.

4.2.1 for the case of weak selection, and complement the results of the local stability analyses carried out in sections 4.3.1 and 4.3.5. The genotypic intrinsic rates of increase, r_{ij} are predictors of the ultimate outcome of selection in the same way as the fitnesses W_{ij} in the discrete-generation case. Except when there is heterozygote superiority or inferiority in the r_{ij}, natural selection tends to fix whichever allele is associated with the highest intrinsic rate of increase when homozygous. When there is heterozygote superiority, Norton showed that the gene frequency either converges to equilibrium or undergoes oscillations in which the gene frequency repeatedly passes through the equilibrium value. The local stability analysis given above can be extended to the continuous-time case by increasing the number of age-classes indefinitely; when combined with Norton's result, it implies that there can be no oscillatory behaviour with heterozygote superiority; if the population converges to a neighbourhood of the equilibrium point, it must eventually reach the equilibrium itself. This result is confirmed by numerous computer calculations of particular cases. The biologically interesting possibility of a stable limit cycle in gene frequency, which was left open by Norton's results, is therefore ruled out for the continuous-time model. No analogue of Norton's results for discrete time has yet been obtained.

4.4.1 *Simplified asymptotic analyses*

Pollak and Kempthorne (1971) gave a similar analysis of the discrete age-class model for some special cases, making the assumption that the *direction* of selection is independent of age, in the sense that for two genotypes A_iA_j and A_lA_m such that $r_{ij} > r_{lm}$, we have $k_{ij}(x) \geqq k_{lm}(x)$ for all x, with the inequality being strict for at least some x. This case was also analysed for the continuous-time model by Charlesworth (1974a), whose approach can easily be modified to deal with discrete age-classes. The method is illustrated here with the case $r_{22} > r_{12} \geqq r_{11}$.

Because of the assumption of infinite effective population size, the frequency of A_2 must lie between 0 and 1 within any finite time-interval, i.e. we have

$$0 \leqq p_2' < p_2(t) < p_2'' \leqq 1,$$

where p_2' and p_2'' are lower and upper bounds for p_2 in the period of time under consideration.

Equation (4.5) gives us for a time t_0

$$B(t_0)p_1(t_0) < \sum_x B(t_0 - x)p_1(t_0 - x)\{k_{12}(x)$$

$$+(1 - p_2'')[k_{11}(x) - k_{12}(x)]\} \qquad (4.58)$$

If $B(t_1)p_1(t_1)$ is the maximum value of $B(t_0 - x)p_1(t_0 - x)$ in the interval $b \leqq x \leqq d$, we have, from equation (4.58)

$$B(t_0)p_1(t_0) \, e^{-(r_{12} + \varepsilon_1)t_0} = (1 - \eta_0)B(t_1)p_1(t_1) \, e^{-(r_{12} + \varepsilon_1)t_1} \qquad (4.59a)$$

where $1 > \eta_0 > 0 \geqq \varepsilon_1$, and where ε_1 satisfies

$$\sum_x e^{-(r_{12} + \varepsilon_1)x}\{k_{12}(x) + (1 - p_2'')[k_{11}(x) - k_{12}(x)]\} = 1 \qquad (4.59b)$$

$B(t_1)p_1(t_1)$ can similarly be related to the maximum value $B(t_2)p_2(t_2)$ of $B(t_1 - x)p_1(t_1 - x)$ in the interval $b \leqq x \leqq d$; in general we can write

$$B(t_i)p_1(t_i) e^{-(r_{12}+\varepsilon_1)t_i} = (1 - \eta_i)B(t_{i+1})p_1(t_{i+1}) e^{-(r_{12}+\varepsilon_1)t_{i+1}}$$

where $t_{i+1} \in [t_i - d, t - b]$, and $\eta_i > 0$.

Extending this process as far back in time as possible, we obtain

$$B(t_0)p_1(t_0) e^{-(r_{12}+\varepsilon_1)t_0} = B(t_n)p_1(t_n) e^{-(r_{12}+\varepsilon_1)t_n} \prod_{i=0}^{n-1} (1 - \eta_i) \qquad (4.60)$$

where $d \leqq t_n \leqq 2d$, and $\eta_i > 0$ for each i. But $B(t_n)p_1(t_n) \exp[-(r_{12} + \varepsilon_1)t_n]$ must be bounded, regardless of the initial gene frequency. If t_0 is chosen sufficiently large, equation (4.60) thus implies that

$$B(t_0)p_1(t_0) e^{-(r_{12}+\varepsilon_1)t_0} < \nu_1$$

where ν_1 is arbitrarily small for arbitrarily large t_0.

A similar procedure can be gone through for allele A_2. We have

$$B(t_0)p_2(t_0) > \sum_x B(t_0 - x)p_2(t_0 - x)\{k_{12}(x) + p'_2[k_{22}(x) - k_{12}(x)]\}$$

so that

$$B(t_0)p_2(t_0) e^{-(r_{12}+\varepsilon_2)t_0}$$
$$= (1 + \eta'_0)B(t_1)p_2(t_1) e^{-(r_{12}+\varepsilon_2)t_1} \qquad (4.61a)$$

where $B(t_1)p_1(t_1)$ is the *minimum* value of $B(t_0 - x)p_2(t_0 - x)$ in the interval $b \leqq x \leqq d$, and where ε_2 satisfies

$$\sum_x e^{-(r_{12}+\varepsilon_2)x}\{k_{12}(x) + p'_2[k_{22}(x) - k_{12}(x)]\} = 1 \qquad (4.61b)$$

and $\eta'_0, \varepsilon_2 > 0$.

By a similar argument to that used for A_1, we can use this to show that, for large t_0

$$B(t_0)p_2(t_0) e^{-(r_{12}+\varepsilon_2)t_0} > 1/\nu_2$$

where ν_2 arbitrarily small for arbitrarily large t_0. Hence, we have

$$\frac{p_1(t_0)}{p_2(t_0)} < \frac{p_1(t_0)B(t_0) e^{-(r_{12}+\varepsilon_1)t_0}}{p_2(t_0)B(t_0) e^{-(r_{12}+\varepsilon_2)t_0}} < \nu_1\nu_2 \qquad (4.62)$$

Expression (4.62) shows that the frequency of A_2 for large t must approach arbitrarily close to one, which is the expected result. The same type of argument can be applied to the case when $r_{22} \geqq r_{12} > r_{11}$, where now $\varepsilon_1 < 0$ and $\varepsilon_2 \geqq 0$, and yields the identical conclusion. As shown by Charlesworth (1974*a*), with heterozygote superiority we can use this type of argument to obtain Norton's result that the gene frequency either approaches the equilibrium or oscillates about it. Pollak and Kempthorne (1971) analysed the case of heterozygote superiority with $k_{11}(x) = k_{22}(x) \leqq k_{12}(x)$, and showed that the gene frequency approaches the equilibrium value of one-half regardless of initial conditions. They also showed that when $k_{11}(x) = k_{22}(x) \geqq k_{12}(x)$, either A_1 or A_2 is fixed, depending on which allele is initially more frequent.

4.5 Selection on a quantitative character

4.5.1 *General considerations*
The results described in the preceding sections of this chapter have been concerned solely with the effects of natural selection on gene frequencies. In this section, we shall consider some aspects of the theory which has recently been developed for dealing with the effects of selection on a quantitative character in an infinitely large age-structured population. The results described here have been developed primarily from the point of view of artificial selection, in which only individuals exceeding some chosen value of the character of interest to the breeder are used for breeding. In the classical, discrete-generation theory of artificial selection (Falconer, 1960, Chapter 11), it is assumed that the character under selection has a known *heritability*, h^2, defined as the proportion of the total variance in the character which is of additive genetic origin. In order to predict the rate of response to selection, we also need to know the *selection differential*, S, which is defined as the deviation from the mean of the unselected population of the mean phenotypic value of the parents chosen for breeding, in the generation under consideration. The *selection response*, R, (which should not be confused with the net reproduction rate) is defined as the deviation of the mean value of their offspring from the mean of the unselected population. If it is assumed that the parents are

chosen independently of each other with respect to their phenotypic values, that linkage disequilibrium and epistatic contributions to the genetic variance are negligible, and that there are no environmental sources of resemblance between parent and offspring, we can write

$$R = h^2 S \tag{4.63}$$

This equation is only valid for one generat:on. If it can be assumed that the heritability and selection differential remain constant in time, repeated application of equation (4.63) enables us to write an equation for the *cumulative selection response*, R_c, which is defined as the deviation of the mean of the character in generation t from its mean in the initial generation. We have

$$R_c = th^2 S \tag{4.64}$$

Provided that an estimate of h^2 is available from prior genetic experiments, this is a result of genuine predictive value. On theoretical and empirical grounds, we would expect h^2 to remain approximately constant for the first 5–10 generations of selection, so that equations (4.63) and (4.64) provide a useful means of predicting the initial rate of response to artificial selection (Falconer, 1960).

Many species of economic importance, such as most domestic animals, have populations with overlapping generations, so that it is not possible to apply equations (4.63) and (4.64) directly. It is therefore of importance to extend the above results to age-structured populations. Some approximate· theory was developed by Dickerson and Hazel (1944) and Rendel and Robertson (1950). An exact matrix approach was introduced by Hill (1974, 1977) and, independently, by Elsen and Macquot (1974). Some refinements have been added by Pollak (1977). The results described below are based mainly on the papers by Hill.

4.5.2 *Prediction of the response to selection*
Breeding values. We shall consider a population described by the discrete age-class model of section 2.3.2, in which the demographic parameters of individuals of both sexes are constant over time, and which is constant in size and age-structure. The size of the population is assumed, however, to be sufficiently large that sampling

effects can be neglected. In order to develop the theory, we will make use of the concept of the additive genetic value or *breeding value* of an individual (Falconer, 1960, pp. 120–2). We let $X(t) = [X(1, t), X(2, t), \ldots X(d, t)]$ be a row vector of the mean breeding values of females aged $1, 2, \ldots d$ at time t, with respect to the character of interest. Similarly, $X^*(t)$ is the corresponding vector of mean breeding values for males aged $1, 2, \ldots d^*$. It is convenient to combine these into a single $(d + d^*)$-dimensional row vector, $\tilde{X}(t) = [X(t), X^*(t)]$ (cf. p. 74), which gives a full description of the state of the population at time t. This scheme is most easily understandable in terms of a character which is measured at a single age, x, such as weight at sexual maturity. A change in the mean breeding value of the individuals of a given sex who are of age x, from one time-interval to the next, must be exactly proportional to the corresponding change in the mean phenotypic value of the individuals in this age–sex class. Changes in the breeding values $X(x, t)$ and $X^*(x, t)$ can therefore be monitored by following the changes in mean phenotypic values of the females and males aged x. If, as is frequently the case, the character can be measured only in one sex, changes in the relevant breeding value can be detected through changes in the mean value of individuals of the appropriate age–sex class. Alternatively, the mean breeding value of individuals of a given class can, in principle, be determined by measurements of the mean value of the progeny produced when they are mated to a standard stock of constant composition. There is thus no theoretical difficulty in assigning breeding values to individuals on which the character of interest cannot be directly measured.

From the relationship of breeding value to the underlying allele frequencies at loci controlling the character (Falconer, 1960, pp. 120–2), it follows that the following expression must hold in the absence of selection

$$\tilde{X}(t) = \tilde{X}(0)G^t \tag{4.65a}$$

where G is the matrix introduced on p. 91. For large t, $\tilde{X}(t)$ is thus given by the asymptotic expression

$$\tilde{X}(t) \sim \tilde{X}(0)A \tag{4.65b}$$

where A is defined on p. 93. From the form of A it follows that the mean breeding value asymptotically approaches the same, constant value for each class of individual, in the absence of selection. This reflects the approach of the underlying gene frequencies to constancy, discussed in section 2.2.1.

Selection differentials with age-structure. Suppose that artificial selection is imposed on the population only at time $t = 0$. The vector of mean breeding values $\tilde{X}(0)$ for the unselected population is incremented by a vector s, so that the vector of mean breeding values of the selected individuals is $\tilde{X}(0) + s$. The components of s depend on the relationship of the character to age, the heritability, and the phenotypic selection differential. Assume, as above, that the character of interest can be measured only at age x. If it can be measured in both sexes, the selection procedure at time $t = 0$ will therefore consist of selecting males and females aged x according to their phenotypes. Only these individuals or their survivors will be used for breeding at the current and future time-intervals. This selection on individuals aged x can be represented by the resulting increments in the mean values of the character, S for females and S^* for males. S and S^* thus define the selection differentials for females and males. If the heritability of the character is h^2 for both sexes, the components of s are given by the following expressions

$$s(x) = h^2 S \tag{4.66a}$$

$$s(d + x) = h^2 S^* \tag{4.66b}$$

$$s(i) = 0 \quad (i \neq x, d + x) \tag{4.66c}$$

If the character can be measured in only one sex, or if selection is practised on one sex alone, all the components of s except that for the appropriate age–sex class are zero.

The response to selection. The response to the selection imposed at $t = 0$, in terms of breeding values, can be found by applying the matrix G to the vector of breeding values of the selected individuals. We thus have

$$\tilde{X}(t) = \tilde{X}(0)G^t + sG^t \tag{4.67}$$

This assumes that the age as well as breeding value of one parent are independent of those of the other. Comparing equations $(4.65a)$ and (4.67), it follows that selection has resulted in an increment sG^t in the breeding value vector. Only part of this is due to the *genetic* effects of selection, since $\tilde{\boldsymbol{X}}(t)$ contains components which have been incremented solely because they relate to individuals who have survived from $t = 0$. In order to distinguish the genetic effects of selection, it is useful to define a matrix $\tilde{\boldsymbol{G}}$ which contains only the survival-related components of $\boldsymbol{G}$, i.e. in which the columns 1 and $d + 1$ have been deleted from $\boldsymbol{G}$. The genetic response to selection at $t = 0$ is thus given by the vector $s(\boldsymbol{G}^t - \tilde{\boldsymbol{G}}^t)$.

If selection is now assumed to be applied at each time-interval, instead of just at $t = 0$, we can describe the cumulative response to selection at an arbitrary time t by a vector of the net increases from $t = 0$ in mean breeding values of each class of individuals, $\boldsymbol{R}_c(t)$. This vector is given by the sum of the responses due to selection in each time-interval prior to t. If selection is practised with the same intensity in each time-interval, and if h^2 remains constant, we obtain the following relation

$$\boldsymbol{R}_c(t) = \sum_{u=1}^{t} s(\boldsymbol{G}^u - \tilde{\boldsymbol{G}}^u) \tag{4.68}$$

For a character which can be measured at age x in both sexes, the phenotypic cumulative response at time t is given by the components $R_c(x, t)$ and $R_c(d + x, t)$, for females and males, respectively.

Equation (4.68) gives an exact prediction of the selection response, subject to the assumptions of the model. Provided that the matrices $\boldsymbol{G}$ and $\tilde{\boldsymbol{G}}$ and the vector s have been specified by the structure of the breeding programme, it is straightforward to use the equation to compute the expected cumulative response at time t. An asymptotic formula can be given for the response at time t, $\boldsymbol{R}(t)$, defined as $\boldsymbol{R}_c(t) - \boldsymbol{R}_c(t-1)$. (This corresponds to the response given by the classical prediction equation (4.63)). We know that $\boldsymbol{G}^t$ approaches the matrix $\boldsymbol{A}$ of p. 93, when t becomes large. Furthermore, when $t \geqq d, d^*$, $\tilde{\boldsymbol{G}}^t$ is a matrix of zero elements, since no individual can survive so long. From equation (4.68), we have

$$\boldsymbol{R}(t) = s(\boldsymbol{G}^t - \tilde{\boldsymbol{G}}^t)$$

so that

$$\boldsymbol{R}(t) \sim s\boldsymbol{A} \tag{4.69}$$

The response per time-interval therefore approaches the same asymptotic value, R say, for each age-class. Using the definition of $\boldsymbol{A}$, this asymptotic response can be written as

$$R = \sum_{i=1}^{d+d^*} s(i)q(i)/2\tilde{T} \tag{4.70a}$$

where the terms $q(i)$ and $\tilde{T}$ are defined on p. 93. In the case of a character which is measured at age x in both sexes, this expression can be combined with equations (4.66) to give

$$R = h^2[Sq(x) + S^*q(d+x)]/2\tilde{T} \tag{4.70b}$$

If the character can be measured in one sex only, the q term corresponding to the other sex must be omitted from this expression.

Hill (1974, 1977) has shown that the asymptotic response is approached fairly quickly. It is therefore tempting to use R to predict the cumulative response to selection by an expression analogous to equation (4.64), i.e. by writing

$$R_c = tR \tag{4.71}$$

As shown by Hill, this provides a good approximation to the cumulative response given by the exact equation (4.68); the difference between the exact and approximate values converges rapidly to a constant level, so that there is little cumulative error in using equations (4.70) and (4.71).

A numerical example. As an example of possible applications of these results, consider the following case of an imaginary herd of cattle, discussed by Hill (1977). The herd is maintained in such a way that all progeny are got by bulls aged two years; one-third each come from cows aged two, three and four years, respectively. The matrix G therefore has the form

$$\begin{bmatrix} 0 & 1 & 0 & 0 & 0 & 0 \\ \frac{1}{6} & 0 & 1 & 0 & \frac{1}{6} & 0 \\ \frac{1}{6} & 0 & 0 & 1 & \frac{1}{6} & 0 \\ \frac{1}{6} & 0 & 0 & 0 & \frac{1}{6} & 0 \\ 0 & 0 & 0 & 0 & 0 & 1 \\ \frac{1}{2} & 0 & 0 & 0 & \frac{1}{2} & 0 \end{bmatrix}$$

The vector q is equal to $(1, 1, \frac{2}{3}, \frac{1}{3}, 1, 1)$, so that the generation time, $\tilde{T}$, is equal to 2.5 years. We assume that selection is practised for weight at one year of age, on bulls only. If weight is normally distributed with standard deviation 40 kg, and the 7 % heaviest bulls are retained for breeding, then standard selection theory tells us that the selection differential on bulls is $S^* = 40 \times 1.9 = 70.6$ (cf. Falconer, 1960, Chapter 11). If the heritability of weight at one year is 0.4, we obtain the following value for R from equation (4.70b)

$$R = (0.4 \times 70.6)/(2 \times 2.5) = 6.08 \text{ kg/year}$$

This expected response can be compared with that for a population in which males are retained for two breeding seasons and females for five, each fertile age-class being assumed to contribute an equal fraction of the progeny. If selection is again practised for weight at 1 year, but now 3.5 % of males and 60 % of females are selected, the selection differentials S and S^* are equal to 24.0 and 88.0, respectively. $\tilde{T}$ is now found to be equal to 3.25 years, and we have $q(1) = q(7) = 1.0$. Equation (4.70b) gives us an expected asymptotic response of

$$R = (0.4 \times 112)/(2 \times 3.25) = 6.89 \text{ kg/year}$$

There is therefore little to choose between the two schemes.

4.6   Conclusions

In this section we bring together the main conclusions from the work described above, and discuss their biological implications. It is convenient to consider density-independent and density-dependent populations separately.

4.6.1 *Density-independent populations*

Fitness measures and their merits. As should be clear from the analyses described above, the theoretical basis for predicting the rate and direction of gene frequency change is less complete than that for studying genetic equilibrium, which was described in the preceding chapter. In particular, the effects of non-random mating with respect to age have not yet been satisfactorily dealt with. When there is random mating between age-classes, and when male and female demographic parameters are the same, it is clear that the intrinsic rate of increase of a genotype, as defined by equation (3.34), is an adequate measure of its fitness for many purposes, provided that the population is in a temporally constant, density-independent environment. The ultimate fate of a new allele at a diallelic locus can be predicted from the relations between the genotypic intrinsic rates, by the results of Norton (1928), Pollak and Kempthorne (1971) and Charlesworth (1974*a*), which are summarised in section 4.4. The rate of spread of a rare dominant or semi-dominant allele is proportional to its effect on the intrinsic rate of increase (sections 4.3.1 and 4.3.2). Finally, with weak selection, equations for gene frequency change can be derived which are similar in form to the first-order difference equations of discrete-generation population genetics, or the Malthusian parameter equations of Fisher (1930, 1941), as described in section 4.2.1.

To make equations (4.11) and (4.15) more comparable with those of discrete-generation population genetics, one can choose the generation-time of a standard genotype, $T_s = \sum e^{-r_s x} k_s(x)$, as a biological unit of time. The equations for Δp_i and $\Delta^2 \ln B$ can be multiplied by T_s and T_s^2 to obtain 'per generation' changes. (A similar operation can obviously be carried out for the corresponding continuous-time model.) The 'per generation' measure of the selective difference between two genotypes with intrinsic rates r_{ij} and r_{lm} can then be taken as $T_s(r_{ij} - r_{lm})$. By the argument used in equation (4.10), we have

$$T_s(r_{ij} - r_{lm}) = \sum_x e^{-r_s x} k_{ij}(x) - \sum_x e^{-r_s x} k_{lm}(x) + O(\varepsilon^2) \qquad (4.72)$$

This is in turn related to equation (3.31) for equilibrium fitness, and is in fact equal to the difference $w_{ij} - w_{lm}$ as given by that equation, plus second-order terms. The equilibrium fitness measure and the intrinsic rate of increase are thus unified by this approximation. When the intrinsic rates of increase of all the genotypes are close to zero, equation (4.72) can be further approximated to give

$$\left(\sum_x x k_s(x)\right)(r_{ij} - r_{lm}) \cong \sum_x k_{ij}(x) - \sum_x k_{lm}(x)$$

If this is used in equation (4.11) to obtain the per generation rate of change of gene frequency, we obtain the type of result that Haldane (1927a) first derived. For weak selection in a slowly growing population, we can conclude that the net reproduction rate of a genotype provides an approximate measure of its fitness, both for predicting the composition of equilibrium populations and for obtaining the rate of change of gene frequency per generation.

If there are demographic differences between the sexes, so that equations (3.12) have to be used as a starting-point, the analysis of pp. 162–3 shows that, with weak selection and random mating with respect to age, the intrinsic rate of increase of a genotype is replaced as its measure of fitness by the mean of the male and female intrinsic rates, the male rate being measured under the demographic conditions of some arbitrary standard population with a stable age-structure. This fitness measure can be translated into a per generation measure by the same type of approximations used above, with the male and female intrinsic rates being multiplied by the male and female 'generation times', respectively. For the case of strong selection, and with possible non-random mating between age-classes, equations (4.28) and (4.29) show how the rate of spread of a non–recessive allele can be determined with demographic differences between the sexes.

The general conclusion is, therefore, that for the case of weak selection and random mating with respect to age, the intrinsic rate of increase of a genotype or, more generally, the mean of the male and female intrinsic rates, provides an adequate measure of fitness in a density-independent and constant environment. This parameter can be used in much the same way as the discrete generation fitness to predict, to a good approximation, the rate of change of frequency and ultimate composition of populations with respect to single loci. Provided that linkage is either very tight or is rather loose, in comparison with the selection intensity, the analysis of equation (4.20) shows that this is true for the two-locus model as well, with some minor qualifications.

If selection is strong, however, there is no unitary measure of fitness. As we have seen (section 4.3.1), the intrinsic rate of increase provides a good predictor of the rate of spread of a rare non-recessive gene, but a different fitness measure must be used for a rare recessive gene. Furthermore, in the neighbourhood of a poly-morphic equilibrium, the intrinsic rate of increase will not predict the rate of change of gene frequency accurately (pp. 181–4), and it also gives an inaccurate prediction of the composition of the equil-ibrium population (section 3.3.3). This conclusion is important for experimenters who wish to study selection by following gene frequency changes in a continuously breeding population such as a *Drosophila* population cage. This has been one of the classical methods for estimating genotypic fitnesses (cf. Du Mouchel and Anderson, 1968). In many cases, particularly in studies of inversion polymorphisms, strong selection has been detected, and some evidence for apparent changes in selection coefficients during the course of selection has been obtained (Dobzhansky and Levene, 1951; Watanabe *et al.*, 1970). The fact that, in age-structured populations, strong selection may lead to deviations from the changes in gene frequency which would be expected on the basis of constant fitness parameters suggests the need for caution in inter-preting apparent evidence for frequency-dependent selection.

One final point concerning strong selection is worth mentioning. With heterozygote superiority in the genotypic intrinsic rates of increase, we have seen that the equilibrium is locally stable (section 4.3.5). The analysis of Norton (1928) left open the possibility that there might be sustained oscillations in gene frequency in this case, since he proved only that there was convergence to a neighbour-hood of the equilibrium, but did not demonstrate convergence to the equilibrium itself. The existence of such oscillations would have been of considerable interest to students of gene frequency changes in fluctuating populations, particularly in view of Chitty's (1960) hypothesis that gene frequency changes may drive cycles of population numbers. As we have seen in section 4.4, oscillations do not appear to be possible in this case, at least as far as the continuous-time model is concerned.

Malthusian parameters and gene frequency equations. Fisher (1930, 1941) introduced equations for gene frequency change in continuous-time models of the form

$$dp'_i/dt = p'_i(m_i - \bar{m}) \tag{4.73}$$

where $m_i - \bar{m}$ is the *average excess* in Malthusian parameters of allele A_i, and p'_i is the frequency of allele A_i among the population as a whole. Fisher appears to have identified the Malthusian parameter of an individual with its intrinsic rate of increase as we have defined it here (see pp. 37–8 of the 1958 edition of *The Genetical Theory of Natural Selection* (Fisher, 1930)). Later users of equations of this form have usually assumed that the Malthusian parameter m_{ij} of a specific genotype A_iA_j is equal to the difference between the *per capita* birth-rate and death-rate for that genotype, and that in principle these can be treated as independent of the genotype-composition and age-composition of the population (e.g. Crow and Kimura, 1970, pp. 190–5). (As we saw in Chapter 1, p. 36, a homogeneous population with the demographic characteristics of A_iA_j will have an asymptotic value for the difference between its *per capita* birth-rate and death-rate which is equal to the intrinsic rate of increase.) It is also usually assumed that the frequencies of the genotypes in the whole population can be approximated adequately by Hardy–Weinberg frequencies, so that m_i and $\bar{m}$ in equation (4.73) can be written as

$$m_i = \sum_j p'_j m_{ij}$$

$$\bar{m} = \sum_i p'_i m_i$$

As mentioned in section 3.1, however, it is not in general possible to assign a fixed birth-rate and death-rate to each genotype in a population whose genetic composition and age-structure are being altered by selection. (Moran, 1962, p. 60; Charlesworth, 1970; Pollak and Kempthorne, 1971). In fact, this can strictly only be done if $m_{ij}(x)$ and $d\ln l_{ij}(x)/dx$ are independent of x for *all* x. This is clearly impossible with the type of reproductive schedule characteristic of most organisms, where reproduction starts some time after birth, so that $m_{ij}(x)$ necessarily depends on x.

Despite these difficulties, it turns out, as we have seen, that an equation similar in form to equation (4.73) can be used to describe gene frequency change in an age-structured population for a continuous-time model with weak selection (equation (4.12)). The gene frequency is measured among new zygotes rather than the whole population. With sufficiently weak selection, this will not make much difference in practice. The use of this equation leads to the conclusion that the rate of change of the population growth-rate under selection is equal to the additive genetic variance in intrinsic rates of increase (the continuous-time analogue of equation (4.14)), which is similar to the Fundamental Theorem that Fisher derived using his Malthusian parameter equation for gene frequency change. Fisher gave no indication, however, that his results were to be restricted to the case of weak selection, together with the other assumptions discussed in Chapter 3; these assumptions are necessary for the basic difference or integral equations for gene frequency and population size to be valid. It is remarkable that equations similar to those proposed by Fisher do, in fact, provide good approximations to the rate of gene frequency change with weak selection. Other interpretations of the Malthusian parameter approach to age-structured models have been proposed by Pollak and Kempthorne (1971), Price (1972), Price and Smith (1972), Nagylaki and Crow (1974) and Nagylaki (1977, pp. 79–88). It is obviously not possible to decide which, if any, of these corresponds to what Fisher had in mind; it is more important to have some firmly-established methods for studying selection in age-structured populations. (Crow (1978) has shown how weighting of individuals by their reproductive value can overcome some of the difficulties of Fisher's approach.)

Furthermore, with temporally-varying environments the Malthusian parameters m_{ij} would be functions of the state of the environment at a given moment, rather than fixed constants; however, equations having the form of equation (4.73) are misleading when regarded as descriptive of gene frequency change in an age-structured population in this situation. Consider for example the two-allele case with A_2 recessive to A_1. Assuming Hardy–Weinberg frequencies, and examining the criterion for A_2 to increase on the average when rare (cf. section 4.3.3), gives the condition

$$E[m_{11}] < E[m_{22}],$$

where E denotes expectation over all environmental states. The criterion for A_1 to increase when rare is the reverse of this, so that one concludes that a polymorphism cannot be maintained by temporal changes in the environment when there is complete dominance, in contrast to the discrete-generation model of Haldane and Jayakar (1963). But we have seen that it is possible to produce examples in which the Haldane–Jayakar model works for age-structured populations, due to the fact that the rate of increase of a rare recessive gene is *not* controlled solely by its effect on the intrinsic rate of increase when selection is strong (section 4.3.3). Another application of equations of the form of equation (4.73) has been to regard them as approximations to discrete-generation difference equations (cf. Kimura, 1955, p. 45). It has been pointed out that, with temporally-varying environments this form of approximation is not adequate (Jensen, 1973; Gillespie, 1973*b*), for rather different reasons from those just discussed for the case of age-structured populations.

4.6.2 *Density-dependent populations*

The results derived in section 4.2.2 show that, with weak selection, the genotypic carrying-capacities as defined by equation (3.40) control the direction of selection in the same way as the intrinsic rates of increase in the density-independent case. In the one-locus case, natural selection tends to maximise the number of individuals in the critical age-group as opposed to the maximisation of the population growth-rate in a density-independent population. It should be borne in mind that this result depends on the assumption that the net reproduction rate of each genotype is a strictly decreasing function of the number of individuals in the critical age-group, and that the maximisation principle applies to the number of individuals in the critical age-group, not necessarily to the population as a whole, or to age-classes outside the critical ones.

The idea that natural selection should, under suitable conditions, result in increasing population density goes back at least as far as Fisher (1930, Chapter 2). Its significance as a general principle was stressed especially by Nicholson (reviewed in Nicholson (1960)). Mathematical treatments for the discrete-generation case have been given by Charlesworth (1971), Roughgarden (1976), Kimura (1978) and Nagylaki (1979*b*), with conclusions very similar to those derived here. Buzzati-Traverso (1955), Beardmore, Dobzhansky and Pavlovsky (1960) and Nicholson (1960), as well as a number of later workers, have described measurements on experimental populations which demonstrate increases in population density or biomass as a result of selection. The experiments of Nicholson are particularly illuminating in this respect. In one case, he kept cage populations of the sheep blowfly *Lucilia cuprina* which were provided with unlimited quantities of food (ground liver) for the larvae, but in which the adults were physically separated from the larvae, and supplied with only a small quantity of food. Since the females require a protein meal in order to lay eggs, the population is essentially limited by competition between females for food, resulting in restricted female fecundity. The results are somewhat complicated by the fact that the population does not settle down to a stationary population size, but undergoes more or less regular oscillations (cf. section 1.4.2). This problem can be overcome by taking the mean population size over a number of days, which behaves fairly regularly (Table 4.2). As can be seen, the number of

Table 4.2. *Increase in population size of an experimental population of* Lucilia cuprina *due to increase in female fecundity*

Period (days)	Mean adult population size
0–88	628
47–146	1161
89–202	2488
147–242	3207
203–298	4281
243–341	6032
299–380	8051
342–434	9348

adults in the population increases steadily over the period of the experiment. It eventually stabilises at a level of about 9300, after 400 days. This increase was shown by Nicholson to be due to the evolution of greatly increased female fecundity under conditions of food scarcity. Since there was no limitation of food supply for the larvae, this increase in fecundity results in an increase in the number of adult flies.

It should be noted that the conclusions about the control of the direction of selection by the genotypic carrying-capacities and the maximisation of population size do not require the different geno-types to respond differently to the density-dependent regulating factors, although this was in fact the case in the example quoted above. As in the discrete-generation case (Charlesworth, 1971), it is perfectly possible for genotypic differences in density-independent components of the life-cycle to result in differences in the carrying-capacities. This may be seen from an example in which density-regulation occurs through a response of the survival of immature stages to the total number of immature individuals at any one time, but where selection affects fecundity. Let the density-dependent probability of survival to reproductive maturity be $w^{(D)}(N)$, and the density-independent net expectation of offspring for an individual who has survived to adulthood be $w_{ij}^{(I)}$ for genotype A_iA_j. The genotypic carrying-capacity, N_{ij}, is defined by

$$\sum_x k_{ij}(x, N) = w^{(D)}(N)w_{ij}^{(I)} = 1$$

From the fact that $w^{(D)}(N)$ is a decreasing function of N, it follows that genotypes with high values of $w_{ij}^{(I)}$ have high carrying-capacities.

This result shows that the often-quoted statement of Haldane (1953), that natural selection will increase the density of a popu-lation only if genotypes differ in their ability to survive a density-dependent source of mortality, is not generally valid. Some experiments of Nicholson (1960) where he artificially adjusted the mortality rates of adult flies in a situation where the size of the population was limited by female fecundity, and found that popu-lation density was inversely related to mortality, provide a model which illustrates this fact.

These conclusions about the maximisation of population size depend on the assumptions of frequency-independent selection and a single-locus genetic system. With discrete generations, it is well known that mean fitness is not necessarily increased when there is frequency-dependent selection, or in multi-locus systems except when linkage is very tight or very loose (reviewed by Turner, 1970). The same applies to population size in the present case. In addition, we have considered only the case of a single-species ecological system. As shown by Levins (1975), Roughgarden (1976) and Léon and Charlesworth (1978) for discrete-generation models, evolutionary change in multi-species systems does not necessarily lead to maximisation of population size. For example, in a predator–prey system where the predator is only weakly regulated by its own numbers, evolutionary change in the predator species tends to increase its own numbers but decreases the numbers of the prey, whereas evolution of the prey species has only a trivial effect on its own numbers. Similar conclusions can easily be derived for age-structured populations by the methods described here for the single-species density-dependent case.

5

The evolution of life-histories

5.1 Introduction

In this chapter we shall discuss the role of natural selection in moulding the forms of life-histories, as described by the relationships of the age-specific survival probabilities and fecundities with age. Theoretical discussion of this topic was initiated by Medawar (1946, 1952), who pointed out that the strength of selection on genes affecting survival or fecundity depends on the ages at which they exert their effects. He suggested that senescence has evolved as a result of stronger selection on genes acting early in life, compared with genes acting late. This idea has since been elaborated by several authors, including Williams (1957), Hamilton (1966), Emlen (1970), and Charlesworth and Williamson (1975). A body of work now exists which provides a firm connection between population genetics theory of the type discussed in the preceding two chapters and biological data on ageing. This will be described in the first part of the chapter.

Another aspect of life-history phenomena which has recently been studied in some detail by evolutionary theorists concerns the relationship of reproductive activity to age. This involves such questions as the effects of selection on the age of reproductive maturity, selection for iteroparity as opposed to semelparity, and whether or not selection can sometimes favour an increase in fecundity with age. This topic was initiated by Cole (1954), who emphasised the advantages of early reproduction for attaining a high rate of population growth. The main features of this and a number of more recent studies will be reviewed in the second section of this chapter. The chapter concludes with a discussion of the relations between theory and observations on life-history phenomena, illustrated with some selected examples.

Before passing on to a detailed account of this work, it may be useful to point out that the viewpoint adopted here is that life-histories evolve as a result of gene frequency changes within populations, under the control of natural selection. The fact that, as we shall see, the theoretical study of life-history evolution frequently reduces to a discussion of the consequences of certain changes in demographic parameters for *population* properties, such as the intrinsic rate of increase or the carrying-capacity, does not mean that *group selection* (Wynne-Edwards, 1962) is being advocated. There are many difficulties in accepting the idea that group selection can often be a significant evolutionary force (Fisher, 1930, p. 49; Maynard Smith, 1964, 1976). The results of the population genetics studies reviewed in the preceding chapter show, however, that, in sufficiently simple circumstances, gene frequency change under natural selection can be understood in terms of the intrinsic rate of increase or the carrying-capacity of the population, so that no appeal to group selection is necessary. The results described in Chapter 4 were mostly based on single-locus models, and it is, of course, likely that life-history evolution involves variability at a great number of loci. Provided that gene effects are approximately additive across loci and that linkage is loose, the results for multi-locus situations can be obtained by simple addition of the contributions to phenotypic change from individual loci (Falconer, 1960; Crow and Kimura, 1970). Single-locus theory thus probably provides a reasonably good guide to the outcome of selection; we are in any case in no worse a situation with respect to life-history evolution than in modelling the effects of selection on any other class of phenotype. For simplicity, the theoretical analysis will be conducted throughout with the discrete age-class model, and sex-differences in demographic parameters will mostly be ignored. As usual, the results can be extended to continuous time by increasing the number of age-classes indefinitely.

5.2 Age-specific gene effects and the evolution of senescence

In this section, we shall use the theory developed in Chapters 3 and 4 to investigate how the selection pressure on a gene is related to the age or set of ages at which it has an effect on the survival or fecundity of its carriers. The relationships which we derive for this purpose are then used to develop some models for the evolution of senescence and related life-history phenomena.

5.2.1 *Selection intensity and age of gene action*

We shall first obtain some expressions for the rate of change in frequency of a gene which has a small effect on survival or fecundity in one age-class only. This model is then generalised to cover a gene with small effects on a number of distinct age-classes. Both density-independent and density-dependent populations will be considered.

Density-independent populations: genes affecting survival. Consider an allele A_2 which is increasing in frequency under selection, in a discrete age-class population with time-independent and density-independent demographic parameters, and which was initially fixed for the alternative allele A_1. If A_2 has only a small effect on survival, we can use equation (4.11) to obtain the following approximate expression for the rate of change per time-interval in the frequency of A_2

$$\Delta p_2 \cong p_2(1-p_2)(r_2-r_1) \qquad (5.1)$$

where $r_1 = p_1 r_{11} + p_2 r_{12}$, and $r_2 = p_1 r_{12} + p_2 r_{22}$. The quantity $r_2 - r_1$ is known as the *average effect of the gene substitution* of A_2 for A_1, on the intrinsic rate of increase (cf. Falconer, 1960, p. 118).

We shall assume initially that both $A_1 A_2$ and $A_2 A_2$ differ, if at all, from $A_1 A_1$ at one age-class alone, say age x. If this is so, the three genotypes $A_1 A_1$, $A_1 A_2$ and $A_2 A_2$ are fully characterised by the survival probabilities $P_{11}(x)$, $P_{12}(x)$ and $P_{22}(x)$, respectively, together with those demographic parameters which are common to all three genotypes. In what follows, it is somewhat more convenient to work with the natural logarithms of these survival probabilities, since independent factors affecting survival are expected to be additive on a logarithmic scale. We can approximate the intrinsic rate of increase for $A_1 A_2$ as follows, using Taylor's theorem

$$r_{12} \cong r_{11} + [\ln P_{12}(x) - \ln P_{11}(x)]\left(\frac{\partial r}{\partial \ln P(x)}\right)_{P_{11}(x)} \qquad (5.2)$$

where r satisfies the standard equation

$$\sum e^{-ry} k(y) = 1 \qquad (5.3)$$

The derivative in equation (5.2) is easily found by applying the rule for the differentiation of an implicit function to equation (5.3):† We have

$$\frac{\partial r}{\partial P(x)} = \sum_{y=x+1}^{d} e^{-ry} k(y) / \left[P(x) \sum_{y=b}^{d} y\, e^{-ry} k(y) \right] \qquad (5.4a)$$

If we write $s(x) = \sum_{y=x+1}^{d} e^{-ry} k(y)$, we obtain the expression

$$\frac{\partial r}{\partial \ln P(x)} = \frac{s(x)}{T} \qquad (5.4b)$$

where T is the familiar measure of generation time, introduced in section 1.3.2. This equation was first given by Hamilton (1966), and has since been discussed by Demetrius (1969), Emlen (1970) and Goodman (1971). It is interesting to note that $s(x)$ is equal to Fisher's reproductive value for age-class $x+1$, multiplied by $e^{-r(x+1)} l(x+1)$, so that $\partial r / \partial \ln P(x)$ is equal to Goodman's eventual reproductive value for age $x+1$, times $e^{-r(x+1)} l(x+1)$. It is evident from equations (5.4) that $s(x)$ is a strictly decreasing function of age, for ages within the reproductive period in an iteroparous species. Furthermore, $s(x)$ takes its maximal value of unity for the pre-reproductive age-classes. As pointed out by Hamilton (1966), these facts have important consequences for the theory of the evolution of senescence, which will be discussed in section 5.2.2.

Equation (5.2) can be rewritten as

$$T_{11}(r_{12} - r_{11}) \cong [\ln P_{12}(x) - \ln P_{11}(x)] s_{11}(x) \qquad (5.5)$$

where T_{11} and $s_{11}(x)$ denote the values of T and $s(x)$ with the demographic parameters of $A_1 A_1$. A similar equation can obviously be written for $r_{22} - r_{11}$ by substituting $P_{22}(x)$ for $P_{12}(x)$. If these relations are substituted into equation (5.1), we can write

$$T_{11} \Delta p_2 \cong p_2 (1 - p_2) \alpha_P s_{11}(x) \qquad (5.6)$$

where α_P is the average effect of the gene substitution of A_2 for A_1

---

† This rule gives $\partial r / \partial P(x) = -[\partial \sum e^{-ry} k(y) / \partial P(y)] / [\partial \sum e^{-ry} k(y) / \partial r]$.

on $\ln P(x)$, such that

$$\alpha_P = p_1 \ln P_{12}(x) + p_2 \ln P_{22}(x) - p_1 \ln P_{11}(x) - p_2 \ln P_{12}(x) \qquad (5.7)$$

It should be noted that, for a given frequency of A_2, α_P is dependent only on the survival probabilities at age x for the three genotypes, whereas $s_{11}(x)$ depends on the age of action of A_2.

Equation (5.6) thus gives the rate of change *per generation* in the frequency of allele A_2, in terms of its effect on the fundamental life-history parameters, α_P, and the scaling factor for its age of action, $s_{11}(x)$. (As selection is weak, both the generation time and the age-of-action factor in equation (5.6) may be taken as equal to the values for any of the three genotypes, without much loss of accuracy, since second-order terms in selective differences can be neglected.) The value of this result is that it allows us to compare gene substitutions at two different loci that affect different ages, but have the same effects on survival probabilities, so that α_P is the same function of gene frequency for both loci. Then whichever gene has the higher value of $s_{11}(x)$ will be selected more strongly, in the sense of changing in frequency more rapidly. From the form of $s(x)$, this means that, if at least one locus acts within the reproductive period, selection will act most strongly on the gene which is expressed earlier. If both genes affect pre-reproductive ages, they will both have the same value of $s_{11}(x)$ (unity), and will be selected at the same rate. Genes which act after the end of reproduction will not be selected. $s(x)$ is closely related to the quantity $q(x)$ used for a similar purpose in Chapters 2 and 4 (pp. 93 and 193).

This result is easily extended to the case of a gene which affects some set, S, of ages in both heterozygous and homozygous states. We have

$$T_{11}(r_{12} - r_{11}) \cong \sum_{x \in S} [\ln P_{12}(x) - \ln P_{11}(x)] s_{11}(x) \qquad (5.8)$$

and a similar expression for r_{22}. If the set S consists of $n + 1$ adjacent ages, $x, x + 1, \ldots x + n$, and the gene has the same effect on survival within each age-class, we obtain the following expression for the rate of change of gene frequency

$$T_{11} \Delta p_2 \cong p_2(1 - p_2)\alpha_P \left[\sum_{i=0}^{n} s_{11}(x + i) \right] \qquad (5.9)$$

where α_P is the average effect of the gene substitution on ln P within each age-class. It is clear from this that, if we compare two different gene substitutions that affect survival by the same amount within each age-class, but which affect two different sets of $n+1$ age-classes, whichever gene initiates its effects earlier in the reproductive period will be selected more strongly.

It is also of interest to consider the effect of selection on a gene which has a positive effect on survival at one stage of the life-history and a negative effect at some other stage. For simplicity, let each of these effects be confined to one age-class, x_1 and x_2, respectively, with average effects of the gene substitution on survival $\alpha_{1P} > 0$ and $\alpha_{2P} < 0$, respectively. We then have

$$T_{11}\Delta p_2 \cong p_2(1-p_2)[\alpha_{1P}s_{11}(x_1) + \alpha_{2P}s_{11}(x_2)] \tag{5.10}$$

Greater weight is placed by the scaling factor on the earlier of the two effects of the gene, provided that at least one of x_1 and x_2 is within the reproductive period. A gene which causes an increase in survival early in life may therefore be selected for even if it has a pleiotropic effect producing decreased survival later on in the reproductive period. This principle is readily extended to genes which express themselves over more than two ages.

Density-independent populations: genes affecting fecundity. Similar calculations can be carried out for the case of a gene which has a small effect on age-specific fecundity at a given age x. Using the methods employed for equations (5.4), we obtain

$$\partial r/\partial m(x) = e^{-rx}l(x)/T = s'(x)/T \tag{5.11}$$

where $s'(x)$ is a scaling factor for the age of action of a gene affecting fecundity, analogous to the factor $s(x)$ for survival. In contrast to $s(x)$, $s'(x)$ is a strictly decreasing function of x if $r \geqq 0$; if r is sufficiently strongly negative, then $s'(x)$ may have a non-monotonic relationship with age, or even be an increasing function of x. This result was first obtained by Hamilton (1966).

If we define α_m as the average effect of the substitution of A_2 for A_1 at a particular diallelic locus, such that $\alpha_m = p_1 m_{12}(x) + p_2 m_{22}(x) - p_1 m_{11}(x) - p_2 m_{12}(x)$, then we have

$$T_{11}\Delta p_2 \cong p_2(1-p_2)\alpha_m s'_{11}(x) \tag{5.12}$$

Hence, if we compare two gene substitutions at different loci with the same effects on fecundity, the gene which acts earlier in the life-cycle will be the more strongly selected, in cases when the initial population (fixed for A_1) has a non-negative intrinsic rate of increase. This result can easily be extended to cover genes which affect more than one age-class, on the lines of equations (5.9) and (5.10). The scaling factor $s'_{11}(x) = e^{-r_{11}x}l_{11}(x)$ reflects the shape of the stable age-distribution for an A_1A_1 population (cf. equation (1.50a)).

Density-dependent populations. We may now consider how these results can be extended to density-dependent populations. Equation (4.23) can be rewritten in the way that yielded equation (5.1), for the case of a single locus with two alleles. We have

$$T_{11}(N_p)\Delta p_2 \cong p_2(1-p_2)[w_2(N_p) - w_1(N_p)] \tag{5.13}$$

where N_p is the equilibrium population density for the current gene frequency, and $w_1(N_p)$ and $w_2(N_p)$ are defined as on p. 167. If A_2 has a small effect on survival, this equation can be approximated by methods similar to those that led to equation (5.6), assuming that A_2 affects survival at a single age x. We can then write

$$w_{12}(N_p) - w_{11}(N_p) \cong [\ln P_{12}(x, N_p) - \ln P_{11}(x, N_p)]\left(\frac{\partial w_{12}(N_p)}{\partial \ln P(x, N_p)}\right)_{P_{11}}$$

where P_{11} and P_{12} are written as functions of N_p to emphasise their potential dependence on density. We note that N_p is equal to N_{11} plus a first-order term in the strength of selection, so that this expression can be approximated as follows

$$w_{12}(N_p) - w_{11}(N_p) \cong [\ln P_{12}(x, N_p) - \ln P_{11}(x, N_p)]s_{11}(x, N_{11}) \tag{5.14}$$

where $s_{11}(x, N_{11}) = \sum_{y=x+1}^{d} k_{11}(y, N_{11})$ is an age-of-action scaling factor comparable with $s_{11}(x)$ for the density-independent case. $s_{11}(x, N_{11})$ is the derivative of the net reproduction rate, R, with respect to the natural logarithm of survival at age x, evaluated at the equilibrium density for an A_1A_1 population. Substituting from equation (5.14) into equation (5.13), and neglecting second-order terms, we obtain the final expression

$$T_{11}(N_{11})\Delta p_2 \cong p_2(1-p_2)\alpha_P s_{11}(x, N_{11}) \tag{5.15}$$

where α_P is the average effect of the gene substitution on survival in a population with density N_{11}, i.e.

$$\alpha_P = p_1 \ln P_{12}(x, N_{11}) + p_2 \ln P_{22}(x, N_{11}) - p_1 \ln P_{11}(x, N_{11})$$
$$- p_2 \ln P_{12}(x, N_{11}).$$

This expression is very similar in form to equation (5.6) for a density-independent population, the only difference being in the restriction that the generation time, age-of-action scaling factor and gene effects on survival are to be measured with the population density set equal to the carrying-capacity for the original A_1A_1 population, instead of the intrinsic rate of increase being set equal to r_{11}. The properties of the scaling factor as a function of age are identical in the two cases. A similar result can be derived for the case of a gene affecting fecundity.

The probability of fixation of a gene with age-specific effects. The above results give some insight into the effects of the age of gene action on the rate of change in frequency of an allele once it has succeeded in establishing itself in a large population. It is, however, of some theoretical interest to ask a different question: what are the differences in the long-term rates of evolution between classes of genes which differ only with respect to their age of action? In order to answer this question, we need to take into account the probabilities of fixation of new, favourable mutations. Let U_i be the probability of fixation, in a large population, of a favourable mutant gene belonging to a given class i, so that a fraction U_i of mutations of this sort which occur at a given time will eventually spread through the population and replace their alternative alleles. Provided that the rates of mutation to favourable alleles are the same in both classes, the relative numbers of favourable mutations of classes i and j which replace their alternative alleles are as $U_i : U_j$ over a time-period which is very long compared with the time taken for an individual gene to spread to fixation (cf. Kimura and Ohta, 1971, p. 12).

In section 4.3.4 we studied the probability of survival of an individual mutant gene in an age-structured population. The results obtained there can be used to relate the U_i to the age of action of the genes concerned. As we saw in Chapter 4, the survival probability for a mutation A_2, introduced into a *stationary* population ($r_{11} = 0$), is proportional to the amount by which it increases the net reproduction rate, R, when heterozygous. If such a mutation also increases survival or fecundity when homozygous, it will spread to fixation if it survives in the population at all. The probability of fixation of a gene which increases survival in a single age-class, x, is thus proportional to $[\ln P_{12}(x) - \ln P_{11}(x)]s_{11}(x)$ (if the population is density dependent, the demographic parameters should be set to their values with the population density equal to N_{11}.) This result can be compared with equations (5.6) and (5.15); it is clear that the relative values of the U_i for genes with the same effects on survival but which act at different ages are determined by the relative values of the scaling factors used for the rate of change in frequency of a favourable gene introduced into a stationary population. A similar argument can be made for genes affecting fecundity.

A somewhat more complex argument is needed when the initial population is increasing in numbers, i.e. when $r_{11} > 0$ (Charlesworth and Williamson, 1975). In such a case, application of the method of Chapter 4 to a gene which has no effect on the demographic parameters of its carriers shows that its chance of survival, U_{11}, is non-zero, since $R_{11} > 1$. This does not mean that the gene becomes fixed, since the branching process method only gives us its chance of surviving random extinction when rare. In practice, a neutral gene that survives extinction simply remains at a low frequency, in an infinitely large population. If we determine the chance of survival, U_{12} say, for a gene which increases the survival or fecundity of A_1A_2 over the values for A_1A_1, the difference $U_{12} - U_{11}$ gives the probability that it survives random extinction and eventually rises to such a high frequency that its fate is controlled by the deterministic equations, and hence goes to fixation. For genes with small effects, $U_{12} - U_{11}$ can be approximated by Taylor's theorem, using equation (4.45) to determine the partial derivatives of U, in the same way that we obtained the derivatives of r earlier in this chapter. The variables used below are defined in section 4.3.4.

Consider first of all genes which affect survival. For pre-reproductive ages $(x < b)$, we have

$$\left(\frac{\partial U}{\partial \ln P(x)}\right)_{P_{11}(x)} = \left\{l_{11}(b) - \sum_{y=b}^{d} [l_{11}(y) \right. $$
$$\left. - l_{11}(y+1)]H_{11}(y, 1-U_{11})\right\}\Big/K \qquad (5.16)$$

where K is a positive constant (Charlesworth and Williamson, 1975). As with the derivative of r, therefore, there is no dependence of this derivative on age of action within the pre-reproductive period.

For later ages, we have

$$\left[\frac{\partial U}{\partial \ln P(x)}\right]_{P_{11}(x)} = \left\{l_{11}(x+1)H_{11}(x, 1-U_{11}) \right. $$
$$\left. - \sum_{y=x+1}^{d} [l_{11}(y) - l_{11}(y+1)]H_{11}(y, 1-U_{11})\right\}\Big/K$$
$$(5.17)$$

This equation implies that the partial derivative of U with respect to survival at age x is a decreasing function of x for $x \geqq b$; the difference between the derivatives for x and $x-1$ is

$$l_{11}(x)H_{11}(x, 1-U_{11})[1 - h_{11}(x, 1-U_{11})]$$

This is always positive for $b \leqq x \leqq d$, since the probability generating function h_{11} is less than unity.

For fecundity, we obtain the expression

$$\left[\frac{\partial U}{\partial m(x)}\right]_{m_{11}(x)} = -\frac{1}{K}\left[\frac{\partial \ln h_{11}(x, 1-U_{11})}{\partial m(x)}\right]$$
$$\times \sum_{y=x}^{d} [l_{11}(y) - l_{11}(y+1)]H_{11}(y, 1-U_{11}) \qquad (5.18)$$

This implies that the relation of the derivative to x must depend in general on the functional form of the offspring distribution for members of a single age-class. When this is a Poisson distribution, we find that (cf. equation (4.46))

$$\left[\frac{\partial U}{\partial m(x)}\right]_{m_{11}(x)} = \frac{U_{11}}{K} \sum_{y=x}^{d} [l_{11}(y) - l_{11}(y+1)] \exp\left[-U_{11}\tilde{M}_{11}(y)\right]$$

$$(5.19)$$

This derivative is a strictly decreasing function of x for both reproductive and pre-reproductive ages, as is the corresponding derivative of r given by equation (5.11).

The qualitative relations with age of the derivatives of r and U are thus very similar. There remains the question of how similar they are quantitatively. This has been investigated numerically by Charlesworth and Williamson (1975), who found that both derivatives behaved similarly. An example is shown in Figure 5.1. This suggests that both the long-term and short-term rates of evolutionary change have similar relationships with the ages of action of genes affecting survival and fecundity.

5.2.2 *The evolution of senescence*

General considerations. We are now in a position to see how these results can be applied to an evolutionary interpretation of senescence and related phenomena. In demographic terms, senescence is defined as the tendency for the age-specific survival probabilities, $P(x)$, (with discrete age-classes) and age-specific fecundities, $m(x)$, to decline with increasing age, for individuals of sufficiently advanced age. With continuous time, senescence with respect to survival is defined as the increase with age in $\mu(x)$, the age-specific death-rate. This process of senescent decline at the level of components of fitness reflects the decline in the performance of many different physiological functions with age (Comfort, 1979). Examples of senescent decline in both fecundity and survival ability are shown in Tables 1.1 and 1.2.

The problem raised by senescence for evolutionists has been to devise a credible theory to account for the establishment and maintenance of this apparently deleterious phenomenon by natural selection. As we saw in the introduction to this chapter, Medawar (1946, 1952) suggested that the key to this problem was the idea that the strength of a selection on a gene whose effects on survival or fecundity are confined to a given age or set of ages is dependent on the age or ages in question, such that the intensity of selection is higher the earlier the time of action of the gene. Medawar proposed that Fisherian reproductive value (equations (1.54)) should be used as an index of the intensity of selection on genes acting at a given age. This proposal was criticised by Hamilton (1966),

Figure 5.1. Partial derivatives of U (dashed lines) and r (full lines) with respect to $m(x)$ (triangles) and $\ln P(x)$ (circles), using vital statistics for the U.S. population of 1939–41 and the Taiwan population of 1906. A Poisson distribution of offspring is assumed for each age-class.

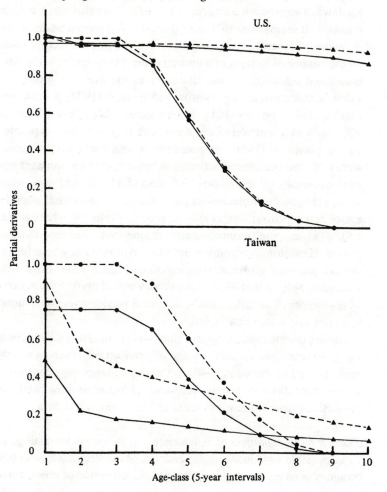

who suggested the use of the partial derivatives of r with respect to age-specific survival and fecundity as indices of selection intensity. These are proportional to the scaling factors $s(x)$ and $s'(x)$ which were discussed above (section 5.2.1). Hamilton justified the use of these derivatives by appealing to Fisher's (1930) formulation of gene frequency change in terms of Malthusian parameters; this lacks, however, a clear derivation from the more basic equations for age-structured populations (section 4.6.1). As we saw above, however, Hamilton's scaling factors for age of action can be given a concrete interpretation in terms of gene frequency change, whereas Medawar's suggestion concerning reproductive value was based on intuition. It seems clear that reproductive value should be discarded in favour of Hamilton's scaling factors or the related factors based on the chance of fixation of a mutant gene. There are, nevertheless, occasional references in the literature to the use of reproductive value in this context, e.g. Wilson and Bossert (1971, p. 124) and Pianka (1974, pp. 98–101). Reproductive value, $v(x)$, behaves differently as a function of age from both $s(x)$ and $s'(x)$, as pointed out by Hamilton (1966). In populations with $r \geqq 0$, $v(x)$ increases with x for the pre-reproductive ages, whereas $s(x)$ is constant and $s'(x)$ decreases (cf. equations (5.4) and (5.11)). In addition, when $r \geqq 0$, $v(x)$ generally reaches its maximum value somewhat after the age of first reproduction, as may be seen in Table 1.4, whereas $s(x)$ and $s'(x)$ decline monotonically throughout the reproductive period. Hamilton also pointed out that, in organisms where fecundity is a geometrically increasing function of age, $v(x)$ *increases* with advancing age, so that Medawar's theory would predict the reverse of senescence. The difference between the two viewpoints is, therefore, not just a technical detail.

Having decided on the appropriate way of measuring the dependence of selection intensity on age of gene action, it seems worthwhile to outline the ways in which this dependence may contribute to the evolution of senescence, and to examine how well the predictions from these ideas seem to fit the biological data.

Paths for the evolution of senescence. It is useful to distinguish three possible ways in which dependence of the strength of selection on age of gene action may contribute to the evolution of senescence.

The discussion will concentrate on senescence in relation to survival; fecundity will be considered in a wider context in section 5.3.

(1) Hamilton (1966) suggested that there would be a higher rate of incorporation of favourable mutations which increase survival at earlier ages within the reproductive period, compared with genes acting later. This would have the effect of gradually raising survival rates at early ages compared with later on. An initially non-senescent life-history, in which survival probabilities are independent of age, would on this view become converted into one which exhibits senescence, in the sense of having lower survival late in life. The above discussion of the relationship between the age of action of a gene and its probability of fixation provides an exact foundation for this idea.

It seems unlikely that this process provides a full explanation of senescence, since so many of the physiological processes involved in senescence seem to be pathological, in the sense of involving a loss of functions which are performed adequately at younger ages. It seems most unlikely that such malfunctioning was characteristic of the younger age-groups of the ancestors of present-day species, and has simply been eliminated by the accumulation of favourable genes. Rather, it seems necessary to invoke processes which involve an accumulation of genes causing positively harmful effects at later ages, as discussed in the following paragraphs.

(2) It is plausible to imagine the existence of genes with pleiotropic effects on survival, such that increased survival at one age is accompanied by decreased survival at another. Equation (5.10) shows that genes of this sort which increase survival early in reproductive life (or in the pre-reproductive period), but decrease it later in the reproductive period, are more likely to spread to fixation than genes which have the reverse effect. The accumulation of such genes would therefore tend to produce senescence, as originally proposed by Williams (1957). A physiological basis for such effects would involve the idea that an increase in the efficiency of some process in youth may entail a cost of some sort, which is manifested later on.

(3) A final contributory factor to senescence may well be the maintenance of deleterious genes by mutation pressure. The

equilibrium frequency of such genes was discussed in Chapter 3 (equation (3.47)). As we have seen (equation (4.72), the fitness measure for the mutant phenotype which appears in the relevant formulae is directly related to the *r* value for the mutant phenotype, so that the same relationship holds between the age of action of a mutant gene and its effect on fitness that we have already discussed in connection with rates of change of gene frequency (section 5.2.1). In the case of a dominant or semi-dominant mutation which has a small effect on survival at a single age *x*, application of the methods of this chapter to equation (3.47*a*) gives the equilibrium frequency of the mutant phenotype as

$$2u/\{[\ln P_{12}(x) - \ln P_{11}(x)]s_{11}(x)\} \qquad (5.20a)$$

Similarly, for a completely recessive gene with the same age of action, the equilibrium phenotype frequency is equal to

$$u/\{[\ln P_{22}(x) - \ln P_{11}(x)]s_{11}(x)\} \qquad (5.20b)$$

We would therefore expect deleterious genes with late ages of action to rise to higher frequencies than genes which exert similar effects on survival early in reproductive life, and therefore to contribute to increasing mortality with advancing age. This idea was apparently first suggested by Medawar (1952). Examples of rare genes causing hereditary diseases with delayed ages of onset are well-known in human genetics, as mentioned in section 3.4.2. The overall contribution to mortality from genes with such drastic effects is probably not great, since they are relatively uncommon. There is good evidence from *Drosophila*, however, that the vast bulk of deleterious mutations have relatively slight effects, and that the per genome rate of mutation is at least 0.25 per generation (Simmons and Crow, 1977). If there is some degree of age-specificity of action among these genes, a significant component of the increase in mortality with increasing age could be attributed to this source.

A related idea was proposed by Haldane (1941, pp. 192–4), who suggested that there might be a significant selection pressure in favour of modifier genes which delay the time of onset of hereditary diseases, since this would lower the impact of the disease on fitness. Medawar (1952) emphasised this process as a major component of the evolution of senescence. While in principle it is clearly capable of operating, it is open to the same objection that Wright (1929*a*, *b*) raised to Fisher's theory of the evolution of dominance (Fisher, 1928*a*, *b*); the selection pressure in favour of a modifier of a rare gene initially at equilibrium under mutation/selection balance must be of the same order as the mutation rate. For example, in the case of a recessive gene, the equilibrium frequency of homozygotes is given by equation (5.20*b*), and the advantage to a rare, unlinked modifier gene which *completely abolishes* its effect can be obtained by multiplying by the denominator, giving a net result of *u*. The advantage to a gene which simply retards the age of onset of the disease is obviously considerably smaller. Such a small effect seems unlikely to be capable of overcoming any deleterious pleiotropic effects of the modifier, or the effects of random sampling of gene frequencies. It thus seems far more plausible to assume simply that mutant alleles have a wide distribution of ages of action, and that those whose effects are restricted to later parts of the life-cycle tend to accumulate at the highest frequencies.

Evidence for age-specificity of gene effects.   Given the existence of age-specific gene effects on survival and fecundity, the evolution of senescence seems almost inevitable by one or other of the mechanisms outlined above. The genetics literature is filled with examples of genes whose action is restricted to particular phases of the life-cycle, and no attempt will be made to review this topic here. (The volume edited by Schneider (1978) reviews various genetic aspects of ageing.) A few selected cases will suffice to demonstrate the general plausibility of postulating age-specificity. We have already discussed evidence for age-specific effects in human hereditary diseases and for genetic involvement in diseases associated with old age (section 3.4.2). In the mouse, several studies have yielded evidence for considerable genetic differences between inbred laboratory strains, both in age-specific mortality patterns and in the incidence and age of onset of individual traits such as circulatory diseases and tumours (Storer, 1966; Festing and Blackmore, 1971). In *Drosophila*, the study of the genetics of mortality differences goes back to the work of Pearl and his associates in the twenties. For example, Gonzalez (1923) showed that flies carrying the mutants *b* and *sp* had significantly higher longevity than wild-type, due to enhanced survival late in life. (This may not have been due to the effects of the mutant genes themselves, but could have been caused by closely-linked genes.) Gowen and Johnson (1946) made careful studies of the life-tables and age-specific fecundity tables of eight different laboratory strains of *Drosophila*, and found considerable strain differences both in longevity and in various aspects of fecundity. Both egg production at the peak and the rate of decline of egg production with advancing age varied significantly, for example.

There is, unfortunately, little evidence for the kind of pleiotropic effects postulated by Williams (1957) and discussed in section (2) above. Very few experiments designed to detect such effects have been carried out. Sokal (1970) compared two strains of *Tribolium castaneum*, the flour-beetle, one of which had been maintained by breeding only from recently eclosed adults; the other was maintained by the usual *Tribolium* culture techniques, which permit adults to live for several weeks. The longevity of the first strain was significantly lower after forty generations than that of the second. This does not, of course, provide conclusive evidence for pleio-tropy, and Sokal interpreted his data in terms of an increased frequency of late-acting mutant genes in the first strain. A similar experiment was carried out by Mertz (1975), who was unable to detect any statistically significant reduction in longevity of *Tribolium* bred exclusively from young adults for about eleven generations. He was able to show that the fecundity of the younger age-classes had been increased significantly by the selection pressure, but that there was no reduction in fecundity at later ages. This seems to provide evidence for age-specific genetic variation in fecundity, but is evidence against pleiotropy. Some recent experiments of this sort with *D. melanogaster* do seem, however, to provide strong evidence of pleiotropy (Rose, 1979; Rose and Charlesworth, in preparation).

Evidence from species comparisons. All three hypothetical paths for the evolution of senescence make similar predictions concerning the effects of the breeding biology and demographic structure of a species on the intensity of selection for senescence. Williams (1957) has reviewed much of the relevant evidence from species comparisons, and the following account summarises his main conclusions. In the first place, senescence is expected to be universally present in organisms in which there is a distinction between the germ plasm and the soma, even if reproduction is asexual, but is not expected in unicellular organisms where no such distinction can be made. This prediction appears to be borne out, as discussed by Williams (1957) and Comfort (1979). The fact that species under natural conditions often appear to show survival probabilities which are independent of age (section 1.2.3) does not mean that senescence is absent in these species. The high probability of death from extrinsic causes will often tend to mask senescent decline, which is observed when individuals are reared in captivity (Comfort, 1979). There is, in any case, some reason to believe that constancy of survival rates is only approximate (Botkin and Miller, 1974), and senescence can sometimes be detected in nature (Caughley, 1966).

Secondly, examination of the forms of the scaling factors for age of action of a gene affecting survival, given by equations (5.4), (5.14) and (5.17), shows that high adult survival rates and low intrinsic rates of increase reduce the differential in favour of younger ages. We would therefore expect lower selection for senescence in stationary populations compared with increasing ones, and in populations where there is a low risk of death from extrinsic causes such as predation, compared with populations in which such risks are high. This prediction correlates well with the differences between groups in rates of senescence. Large mammal species tend to have higher maximum life-spans than small ones (Comfort, 1979), and show considerably higher survival rates under natural conditions (Caughley, 1966). Similarly, birds tend to have higher adult survival rates than do mammals of similar size, presumably because of lower vulnerability to predation; they also show greater maximum life-spans in captivity. It is in groups that have extremely high levels of unavoidable mortality, such as small insects, that high rates of senescence are expected and observed. The observation of Sacher (1959, 1978) that the maximum life-span of mammalian species is correlated with the ratio of brain weight to body weight can also be interpreted on these lines; one can hypothesise that such a correlation reflects a relation between brain size and intelligent behaviour, the latter being correlated with a lower susceptibility to extrinsic causes of death.

Thirdly, species where fecundity increases with age, such as many cold-blooded vertebrates (section 1.2.3), should show relatively low rates of senescence with respect to survival, since increasing fecundity tends to reduce the differential in favour of younger age-classes that is seen in the age-of-action scaling factors. This is in agreement with the observation that species of large reptiles, amphibia and fish may have very high maximum life-spans (Williams, 1957; Comfort, 1979).

Fourthly, the theory suggests that senescence with respect to survival should start immediately after the age of first reproduction, since the scaling factor $s(x)$ is constant for all pre-reproductive ages, and then starts to decline. Owing to the inadequacies of the data on animal life-tables, and the confounding effects of differences in ecology between juvenile and adult stages in many species, it is difficult to be certain how generally this prediction is verified. Human life-tables do show a minimum in mortality rates at or near the age of puberty, with a continuous increase with age after that point (see Williams, 1957; Hamilton, 1966, and Table 1.2); the rise in mortality after puberty fits the theory well. This raises the question of why juvenile mortality rates should be so high in many species. On the present theory, there should be no selection for changes in the distribution of mortality within the pre-reproductive period, except insofar as any age-specific gene effects overlap the beginning of reproduction; this might lead to some tendency for higher survival rates *early* in the pre-reproductive period. It is easy to invoke developmental and ecological reasons for high juvenile mortality, such as the small size of juveniles and consequent vulnerability to predation and accidental death. In vertebrates, the existence of a period of immunological tolerance in very young individuals necessarily implies a relatively high rate of death due to infectious disease.

Finally, there should be no opposition to the spread of genes which reduce survival after the age of termination of reproduction, in species where there is an abrupt termination. This does not, of course, imply that the end of reproduction should necessarily be followed by immediate death, merely that selection against senescent decline is ineffective after reproduction has ceased. The long post-menopausal life of women in contemporary societies may have no selective relevance, since in primitive conditions only a small fraction probably survived to the age of menopause. Possible selective advantages of an extended post-menopausal life in terms of greater ability to care for children or grand-children have, however, been discussed by Williams (1957) and Hamilton (1966). This pre-supposes the post-menopausal survival of at least some women.

5.3 The evolution of reproductive patterns in relation to age

This section is concerned with exploring the consequences of selection on the pattern of reproduction in relation to age. This is a topic which has attracted considerable attention from evolutionary theorists, especially in the last decade; space does not permit detailed discussion here of all the ideas which have been put forward in this area. Instead, we shall concentrate on certain key concepts which seem to have emerged. We begin by using some of the theory developed in the preceding section to illustrate the idea that selection frequently favours early reproductive maturity and high levels of reproduction early in reproductive life. We go on to discuss various factors which tend to work in the opposite direction, and that favour such traits as iteroparity versus semelparity and deferred reproduction versus early reproduction. The idea that life-histories have evolved under selection for an optimal allocation of resources at each age between growth, survival and reproduction is then discussed.

5.3.1 *Selection and time of breeding*

Following the seminal paper of Cole (1954), the existence of a selection pressure in favour of early breeding has been emphasised by numerous authors, e.g. Lewontin (1965), Hamilton (1966), MacArthur and Wilson (1967, Chapter 4), Charnov and Schaffer (1973). This selection pressure can be viewed mathematically in several different ways, as may be seen from the works just cited. We shall confine ourselves here to examining just two of these viewpoints, relating to the higher intensity of selection for increased fecundity early in life, and to the relative advantages of semelparity and iteroparity, respectively.

Selection for increased fecundity in relation to age. The mathematical basis for this has already been developed in equations (5.11) and (5.19) above, and the evolutionary consequences of these equations have already been reviewed. We saw that the earlier the point in the life-cycle at which a gene increasing fecundity acts, the more strongly it will be selected, provided that the population is either stationary or increasing in size. The higher the mortality in the population and the higher its intrinsic rate of increase, the faster the scaling factor $s'(x)$ in equation (5.11) falls off with age, and so the higher the differential selection in favour of increasing fecundity early in life. On this basis, the types of model proposed above for the evolution of senescence predict that, once reproduction has been initiated, it should rise to a maximum early in reproductive life, and then should decline slowly in a fashion that roughly parallels the decline in $s'(x)$ with age.

This type of reproductive pattern is characteristic of many iteroparous species, and is illustrated by the human and *Drosophila* examples of section 1.2.3. As mentioned in that chapter, there are, however, other reproductive patterns which must be accounted for. One of these is the constancy of clutch size or litter size over age seen in many species of birds and mammals. There is evidence in birds that the mean size of the clutch in many species is considerably lower than the maximum size which is physiologically possible, and has been adjusted by selection close to a size which enables the maximum number of young to be reared by the parents (Lack, 1954, 1966). Senescent decline affecting the physiological limit to egg production would not, therefore, appreciably affect the number of eggs laid, until very late in the life-cycle. A decline with age in the ability of parents to care for their young might, however, be expected (cf. p. 5). Similar principles probably apply to the mammalian examples. The other outstanding exception to the principle of a decline in fecundity with age is when reproductively mature individuals continue to grow in size, resulting in an increase in fecundity with age (section 1.2.3). Any decline in fecundity with age may thus be masked, although individuals of sufficiently advanced age may show senescent decline (Gerking, 1959). Evolutionary factors which favour an increase in fecundity with age are discussed in section 5.4.2.

A corollary of this differential selection pressure in favour of early fecundity is that selection should tend to minimise the age of first reproduction (b), as far as is consistent with any harmful effects of premature reproduction. This idea was stressed by Lewontin (1965), who showed by numerical methods that a small reduction in b may contribute as large an increase in fitness as a substantial increase in fecundity. There is some evidence from *Drosophila* that development time may be near its lower physiological limit. For example, Clarke, Maynard Smith and Sondhi (1961) obtained a strongly asymmetrical response to artificial selection for development time in *D. subobscura*, with a realised heritability of 0.06 for downward selection and 0.19 for upward selection. It should be noted that Lewontin's calculations were based on the assumption of extremely high r values, since he was concerned with modelling selection in a colonising situation. As stressed by Mertz (1971a, b), species with low or negative rates of increase and low mortality will not be subject to such intense selection for early reproduction. Selection in a declining population with low mortality may actually favour delayed reproduction, since $s'(x)$ can be an *increasing* function of x for at least part of its range under these conditions. (A population obviously cannot continue to decline in numbers over the period of time needed for evolutionary change without going extinct, so that this point is probably somewhat academic.) There seems to be a general correlation between low mortality and population growth-rates and deferred reproduction in many species (Lack, 1966; Mertz, 1971a, b; Goodman, 1974) which fits well with the effects of these demographic factors on $s'(x)$. A full discussion of the selective control of the age of reproductive maturity must take into account the possible adverse effects of reproduction on subsequent growth and survival; this topic will be considered in section 5.3.2.

Iteroparity versus semelparity. The pressure of selection in favour of early breeding can also be examined by the method originally proposed by Cole (1954), and later modified by Gadgil and Bossert (1970), Bryant (1971), and Charnov and Schaffer (1973). Other things being equal, it seems from the above considerations that selection should tend to favour reproduction as early as possible in the life-cycle, so that in species with seasonal breeding, the life-history associated with highest fitness is one in which reproduction is limited to the first breeding season. This is, of course, the reproductive pattern characteristic of annual plants and univoltine insects and is extremely widespread. One limitation to this mode of reproduction is the necessity of having a high level of reproduction in the single breeding season, in order to compensate for the loss of progeny from later seasons. There may well be physiological upper limits to fecundity; in organisms with parental care, there may also be a selective limitation to fecundity due to the need to have a clutch of sufficiently small size to be cared for adequately (p. 224).

The strength of this limitation can be assessed as follows, using the method of Charnov and Schaffer (1973), which is the most general formulation among those cited above. Consider an iteroparous life-history with constant adult survival, P, from one age-class to the next, and constant age-specific fecundity, m_i. Let the age of first reproduction be b, and the probability of survival to this age be $l(b)$. Then, if there is no upper age limit to survival or reproduction, the characteristic equation for the intrinsic rate of increase, r_i, can be written as follows (cf. p. 61)

$$1 = l(b)m_i(e^{-r_ib} + P e^{-r_i(b+1)} + P^2 e^{-r_i(b+2)} + \ldots)$$

i.e.

$$1 = [l(b)m_i e^{-r_ib}]/(1 - P e^{-r_i}) \tag{5.21a}$$

A semelparous life-history with age of reproduction equal to b and probability of survival to age b equal to $l(b)$, but with fecundity equal to m_s, has an intrinsic rate of increase, r_s, given by

$$1 = l(b)m_s e^{-r_sb} \tag{5.21b}$$

For equal fitness of the two life-histories, we require $r_i = r_s = r$, say; the condition for this is

$$m_i/m_s = (1 - P e^{-r}) \tag{5.22}$$

For a semelparous life-history to be at an advantage over iteroparity, m_s must exceed the value given by this equation. The higher P and the lower r, the greater the value of m_s required. Iteroparity is therefore favoured by high P and low r. If $P = \frac{1}{2}$ and $r = 0$, fecundity would have to be doubled for semelparity to do just as well as iteroparity; if $P = 0.9$ and $r = 0$, fecundity would have to be increased tenfold.

This result is particularly clear when $b = 1$, so that reproduction is initiated after one season. Equations (5.21) yield the following expressions for r_i and r_s

$$e^{r_i} = l(1)m_i + P \qquad (5.23a)$$

$$e^{r_s} = l(1)m_s \qquad (5.23b)$$

For equality of the intrinsic rates, we therefore need

$$m_s = m_i + [P/l(1)] \qquad (5.24)$$

High adult survival relative to juvenile survival is thus particularly favourable to iteroparity in this case. This accords well with the observation that semelparity with $b = 1$ is particularly common among insects of the temperate zone, and in plants which are associated with unstable, temporary habitats (Stebbins, 1950, Chapter 5). In both of these cases, the chance that an adult survives from one year to the next is presumably very low. These results are more fully discussed by Schaffer and Gadgil (1975).

The conclusions can readily be extended to density-dependent populations. Consider, for example, the case when survival to reproductive maturity is a strictly decreasing function of the number of individuals in the critical age-group, $l(b, N)$ (cf. section 1.4.2). The following equations correspond to equations (5.21), and give the carrying-capacity, $\hat{N}$, for iteroparous and semelparous populations respectively

$$1 = l(b, \hat{N})m_i/(1 - P) \qquad (5.25a)$$

$$1 = l(b, \hat{N})m_s \qquad (5.25b)$$

From the results of Chapter 4 (section 4.2.2), we can use the carrying-capacity as a measure of fitness, so that iteroparous and semelparous life-histories confer equal fitness when

$$m_i/m_s = (1 - P) \qquad (5.26)$$

This result is equivalent to equation (5.22) for the density-independent case.

The effects of temporally varying environments. Murphy (1968) suggested that temporal variation in the survival probabilities of juveniles might create a selective advantage for iteroparity. Consider, for example, the extreme case of an environment in which there is an occasional season in which juveniles completely fail to survive. A semelparous life-history would be lethal in this season, whereas an iteroparous life-history would be able to produce offspring in the succeeding, favourable season, as a result of the survival of adults. Less extreme variation in the environment would similarly tend to promote iteroparity, as Murphy showed by means of computer calculations of the spread of a gene causing iteroparity (see also Hairston, Tinkle and Wilbur, 1970). This principle can be illustrated with the following example. Suppose that the environment alternates between two equally frequent states, such that the probability of survival of individuals to age 1 is $(1+s)l(1)$ or $(1-s)l(1)$, with a mean of $l(1)$. Provided that the environmental states alternate at intervals which are very long compared with the life-span of an individual, so that age-structure is mostly constant, the asymptotic mean rate of increase of a population is equal to the mean of the intrinsic rates for the two environments (cf. p. 176). Using equation (5.23a), the mean rate, $\bar{r}_i$, for an iteroparous population which starts breeding at age 1 is given by the following equation

$$\bar{r}_i = \tfrac{1}{2} \ln \left[l(1)(1+s)m_i + P \right]\left[l(1)(1-s)m_i + P \right]$$
$$= \tfrac{1}{2} \ln \left[l^2(1)(1-s^2)m_i^2 + 2l(1)m_iP + P^2 \right] \qquad (5.27a)$$

Similarly, the mean rate for a semelparous life-history, $\bar{r}_s$, is given by the equation

$$\bar{r}_s = \tfrac{1}{2} \ln \left[l^2(1)(1-s^2)m_s^2 \right] \qquad (5.27b)$$

As we saw in section 4.3.3, the criterion for the spread of a rare, non-recessive gene in a temporally varying environment of this sort is that the mean rate of increase for the heterozygote exceeds that for the initial homozygotes. For a gene causing semelparity to be capable of invading an iteroparous population, equations (5.27) therefore imply that m_s must exceed the value given by the following equation

$$m_s^2 = m_i^2 + \frac{P[2l(1)m_i + P]}{l^2(1)(1 - s^2)} \tag{5.28}$$

Clearly, the higher s the greater must be the fecundity of the semelparous genotype in order for it to spread. A similar result holds if the fecundities m_i and m_s are subject to environmental fluctuation. On the other hand, if the adult survival alternates between $P(1 + s)$ and $P(1 - s)$, the equivalent of equation (5.28) is

$$m_s^2 = m_i^2 + \frac{P[2l(1)m_i + P(1 - s^2)]}{l^2(1)} \tag{5.29}$$

Fluctuations in adult survival thus tend to favour semelparity.

The weakness of these results in that they are strictly valid only for the case when the population spends a long time in each environment. As discussed in section 4.3.3, no adequate theory has yet been developed for the dynamics of selection in age-structured populations, when the time period of the environmental changes is short compared with the time needed to stabilise age-structure. The same difficulty applies to Schaffer's (1974b) treatment of the consequences of environmental fluctuations for the reproductive effort model discussed on pp. 237–52 below. To what extent the conclusions arrived at above are affected by the time-scale of environmental change is a matter for future theoretical research. Empirical evidence for correlations between iteroparity and variable juvenile survival or between semelparity and variable adult survival is scanty (Murphy, 1968; Stearns, 1976, 1977).

Synchronisation of breeding in semelparous species. A life-history phenomenon of considerable interest is the tendency in some semelparous groups for synchrony of breeding of all the members of the species in a given locality, so that adults are only produced once every b years, as in the periodical cicada (p. 29). Several different explanations of this synchrony have been put forward. Lloyd and Dybas (1966a, b) suggested that it could be an evolutionary response to predation pressure; they pointed out that the large numbers of breeding individuals which appear every b years are likely to swamp the capacity of predators for destruction, so that members of a synchronised population are likely to have a higher rate of survival than members of an unsynchronised population. The difficulty with this view is that it is hard to see how there could be any advantage to a variant life-history with synchronisation, when introduced at a low frequency into an unsynchronised population, so that one would have to appeal to group selection as an evolutionary mechanism.

It therefore seems more fruitful to seek a purely demographic model in order to explain synchrony. Hoppensteadt and Keller (1976) constructed a model in which there was assumed to be predation on the adults of the semelparous species, with a threshold level for the number of adults, above which predation was ineffective in removing individuals. The value of the threshold reflects the number of predators, and was assumed to be positively related to the number of adults in the preceding year. The population of juveniles of the semelparous species was assumed to be regulated by some fixed upper limit to the total number of juveniles; juvenile survival was density-independent below this limit. This model is capable of producing situations in which only synchronised populations are ecologically stable, given suitable choices of parameters.

A different model was studied by Bulmer (1977), who assumed that the primary ecological factor was competition between juveniles. He suggested that such competition might be more severe between than within age-classes, since larger (older) individuals are likely to be more successful in competition than smaller (younger) ones, leading to a kind of competitive exclusion such that the population is dominated by one age-group. His analytical and numerical results confirmed that this kind of asymmetrical competition is capable of generating a situation in which only a synchronised population is stable. Bulmer also investigated the consequences of predation and parasitism, and concluded that they could be important in reinforcing any tendency to synchrony caused by competition, but could not produce synchrony by themselves. His predation models had a very different structure from that of Hoppensteadt and Keller (1976), so that there is no contradiction between their results. Discrimination between these alternative explanations will clearly require further empirical studies.

5.3.2 Optimal life-histories

Up to now, our discussion of reproductive patterns has neglected the possible existence of constraints on changes in survival and fecundity at a given age, other than those imposed by physiological limits to the intensity of reproduction. There has recently been considerable interest in the theoretical study of life-history models which assume various types of constraints on the relationships between survival and fecundity, e.g. that survival and fecundity at a given age are inversely related (Gadgil and Bossert, 1970; Fagen, 1972; Schaffer, 1974*a*, *b*; Taylor, Gourley and Lawrence, 1974; Schaffer and Gadgil, 1975; Charlesworth and León, 1976, León, 1976; Schaffer and Rosenzweig, 1977; Michod, 1979). The principle common to these studies is the idea that natural selection in a density-independent environment should establish a life-history which corresponds to a maximum in *r*, subject to the constraints assumed in the model.

Why maximise r? Before proceeding further, it is useful to consider briefly the justification for treating life-history evolution as a problem in maximising *r*. We can use for this purpose Maynard Smith's (1972) concept of an *evolutionarily stable strategy* or ESS. A population is said to be at an ESS, with respect to some phenotype or set of phenotypes, if a rare gene introduced into the population, and whose carriers have a different phenotype from the mean of the population, is always eliminated. This provides a necessary condition for the population to be a stable point with respect to evolutionary change. If the population is not at an ESS, at least some classes of mutations with altered phenotypes will be capable of spreading, with a consequent change in the mean phenotype of the population. The application of this idea to life-history evolution is obvious; in a density-independent environment, a maximum in *r* must correspond to an ESS, since a mutation which decreases *r* is eliminated by selection (cf. sections 4.3.1 and 4.3.2). It is important to distinguish between *local* and *global* maxima in *r*. It is perfectly possible in principle for there to be several points in the space specified by the permissible sets of phenotypes which are local maxima in *r*; each of these corresponds to an ESS with respect to mutations of sufficiently small effect. Only one of these can be a global maximum in r, unless several turn out to have equal values of r. Some examples of the coexistence of maxima in r are discussed by Schaffer and Rosenzweig (1977). It may be difficult for a population which, for historical reasons, is located at an ESS which is not a global maximum in r to approach the global maximum, unless a mutation of sufficiently large effect occurs. A complete mathematical solution of the life-history problem therefore requires an enumeration of all the life-histories which correspond to local maxima in r, not just the global maximum. This can rarely be achieved.

It should be pointed out that the ESS analysis ought strictly to be conducted by means of calculations of the survival probabilities of mutant genes, rather than by the use of deterministic conditions for their spread, using the methods described earlier (pp. 211–14). As shown by Charlesworth and Williamson (1975), the survival probability of a rare gene is generally closely related to its effect on r, so that attention will be restricted here to the simpler calculations involving r.

The technicalities of maximising r are simplified to some extent by the following facts, discussed in more detail by Taylor *et al.* (1974). If a life-history corresponding to a certain reproductive function $k'(x)$ yields a local maximum in r, r' say, the characteristic equation (1.45) gives the relation

$$\sum e^{-r'x}k'(x) = 1 \qquad (5.30a)$$

Any small perturbation to another reproductive function $k''(x) \neq k'(x)$ with, by definition, $r'' < r'$, must therefore be such that

$$\sum e^{-r'x}k''(x) < \sum e^{-r'x}k'(x) = 1 \qquad (5.30b)$$

The sum $\sum e^{-r'x}k(x)$ is therefore at a local maximum when $k(x) = k'(x)$. Conversely, if $\sum e^{-r'x}k(x)$ is at a local maximum when $k(x) = k'(x)$, r' is a local maximum in r. This can be summed up by saying that the necessary and sufficient conditions for a life-history $k'(x)$ to be at a local maximum in r is that the sum $\sum e^{-r'x}k(x)$ is at a maximum when $k(x) = k'(x)$, with r' being specified by equation (5.30a). The conditions for a global maximum in r can be expressed similarly.

Analogous principles can be developed for studying life-history evolution in a density-dependent environment (Charlesworth and

León, 1976). Using the model of density-dependence developed in section 1.4.2, a given life history $k'(x)$ is associated with a carrying-capacity $\hat{N}'$, determined by the equation

$$w'(\hat{N}') = \sum_x k'(x, \hat{N}') = 1 \tag{5.31a}$$

If $w(N)$ is a strictly decreasing function of N, it follows from the results described in section 4.3.2 that a mutant gene associated with a perturbed life history $k''(x)$ will be eliminated if and only if

$$w''(\hat{N}') = \sum_x k''(x, \hat{N}') < w'(\hat{N}') = 1 \tag{5.31b}$$

This is, therefore, the necessary and sufficient condition for the life-history $k'(x)$ to be an ESS. It is also the condition for the carrying-capacity $\hat{N}'$ to be a maximum. There is thus no real difference between the density-dependent and density-independent cases, as far as the techniques of determining the ESS life-histories are concerned.

Optimisation of the age of reproductive maturity. In order to illustrate these ideas, consider the problem of determining the optimal age of first reproduction, a question which has been discussed in various contexts by Gadgil and Bossert (1970), Wiley (1974a, b) and Bell (1976). We shall use the type of model suggested by Schaffer (1974a), which assumes that fecundity is an increasing function of size, and that growth is a multiplicative process such that size at age x, $G(x)$ say, equals $g(x-1)G(x-1)$. The *growth-rate* of individuals aged x is thus $g(x)$. The size of a zygote is equal to $G(0)$, so that we have

$$G(x) = G(0) \prod_{y=1}^{x} g(x-y) \tag{5.32}$$

This is an expression of the same form as that for $l(x)$ in equation (1.15), and it is convenient to combine growth and survival into one formulation by writing

$$\tilde{P}(x) = P(x)g(x) \tag{5.33a}$$

$$\tilde{l}(x) = \prod_{y=1}^{x} \tilde{P}(x-y) \qquad (5.33b)$$

($\tilde{l}(x)$ as defined here should not be confused with the $\tilde{l}(x)$ in equation (1.18), which was defined for an entirely different purpose). Finally, we can represent fecundity at age x in terms of the fecundity per unit weight, $\tilde{m}(x)$, so that

$$m(x) = G(x)\tilde{m}(x) \qquad (5.34a)$$

$$k(x) = \tilde{l}(x)\tilde{m}(x) \qquad (5.34b)$$

In cases when fecundity is not dependent on size but on some other function of age, such as experience in catching prey (Orians, 1969), we can use the same type of representation, except that $G(x)$ represents growth in reproductive efficiency rather than size.

The problem is to find what value of the age of reproductive maturity, b, yields the fittest life-history, taking into account the advantage of postponing reproduction until the individual has had time to grow to a size where its fecundity is high, balanced against the risk of death before reaching age b. Consider first the case of a semelparous organism in a density-independent environment. The possible life-histories which are to be compared are characterised by the values of $\tilde{l}(x)$ for $x = 1, 2, \ldots b$, and the fecundity per unit weight at age b, $\tilde{m}_s$. Both $\tilde{m}_s$ and $l(x)$ for $x \leq b$ are assumed to be independent of b, and are determined purely by some specified model for growth, survival and reproduction. The equivalent of equation (5.21b) is thus

$$1 = \tilde{l}(b)m_s\, e^{-r_s b} \qquad (5.35)$$

For $r > 0$, e^{-rb} is a decreasing function of b. Using the argument outlined above (p. 232) for finding a maximum in r, it follows that a value of $b > 1$ can correspond to a maximum in r_s only if $\tilde{l}(x)$ has a maximum at some value of $x > 1$. Even if this condition is satisfied, r_s may be so high, if $\tilde{m}_s$ is large, that $e^{-r_s b}$ falls off so rapidly with increasing b that $b = 1$ gives the maximum value for r_s. Now $\tilde{l}(x)$ is composed of two components, $l(x)$ and $G(x)$. The first of these decreases with increasing x as a result of mortality, and the other increases as a result of growth (see Figure 5.2). On general biological grounds, it is reasonable to assume that $G(x)$ would increase rapidly with age and then level off. For $\tilde{l}(x)$ to have a maximum with $x > 1$, therefore, $l(x)$ must decrease sufficiently slowly, as x is increased above unity, that its decline is outweighed by the increase in $G(x)$. The position of the maximum in $\tilde{l}(x)$ must be to the left of the point at which growth ceases (Figure 5.2).

With sufficiently small $\tilde{m}_s$ and sufficiently high survival at age 1 and above, the life-history with maximum r_s will have $b > 1$. If $\tilde{m}_s$ is high and survival is low, then the optimum life-history will reproduce at age 1, reflecting the higher intensity of selection for early breeding under such conditions (section 5.3.1).

These considerations can readily be extended to density-dependent populations, using the method described above (p. 233). Consider first the case of a population regulated solely by dependence of fecundity on density, so that $\tilde{m}_s$ can be written as a decreasing function of the number of adults, $\tilde{m}_s = \tilde{m}_s^{(I)} \tilde{m}_s^{(D)}(N)$ (cf. equation (1.83a)). The carrying-capacity, $\hat{N}$, is given by the equation

$$1 = \tilde{l}(b) \tilde{m}_s^{(I)} \tilde{m}_s^{(D)}(\hat{N}) \tag{5.36}$$

The value of b which maximises $\hat{N}$ can be obtained by the same method as above. In this case, there is no factor corresponding to

Figure 5.2. The graphs of $l(x)$ (solid circles), $G(x)$ (open circles) and $\tilde{l}(x)$ (crosses) for an imaginary example. $G(x)$ and $\tilde{l}(x)$ are in arbitrary units.

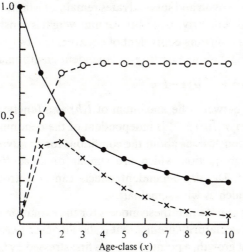

$e^{-r_s b}$, so that the optimal value of b corresponds with the value which maximises $\tilde{l}(x)$. It is therefore easier than in the density-independent case to satisfy the conditions for postponed reproduction to be optimal. A similar conclusion is obtained if density-dependence is restricted to survival from age 0 to age 1. If, however, density-dependence is mediated through a survival factor which is applied equally to each age class, we can write

$$\tilde{l}(x) = \tilde{l}^{(I)}(x)P_D^x(N) \tag{5.37}$$

where $P_D(N)$ is the density-dependent component of survival, such that $P_D(0) = 1$, and $\tilde{l}^{(I)}(x)$ is density-independent. The equation for the carrying-capacity $\hat{N}$ is now

$$1 = \tilde{l}^{(I)}(b)\tilde{m}_s P_D^b(\hat{N}) \tag{5.38}$$

The factor $P_D(\hat{N})$ plays a similar role to e^{-r_s} in the density-independent case described by equation (5.35). A high value of $\tilde{m}_s$ implies that $\hat{N}$ must be high, so that $P_D^x(\hat{N})$ falls off rapidly with age, and accentuates the advantage of early reproduction. Density-dependent regulation of this type does not, therefore, particularly favour postponement of reproduction.

The case of an iteroparous species can be analysed in a very similar way. Assume for simplicity that after reproductive maturity has been reached, growth and survival rates remain constant, so that $\tilde{P}(x) = \tilde{P}$, for $x \geq b$. Similarly, fecundity per unit weight is constant at $\tilde{m}_i$. We obtain the following equivalent of equation (5.21a), for the intrinsic rate of increase in a density-independent environment

$$1 = [\tilde{l}(b)\tilde{m}_i \, e^{-r_i b}]/(1 - \tilde{P} \, e^{-r_i}) \tag{5.39}$$

The relationship between the maximum of $\tilde{l}(b)$ as a function of b and the maximum of $\tilde{l}(b) \, e^{-r_i b}$ is independent of the denominator, so that we reach conclusions about the conditions for an advantage to postponed reproduction which are similar to those for the semelparous case. Density-dependent effects can be introduced into this model much as with semelparity.

It is interesting to note that these models for the evolution of the age of reproductive maturity predict that it should never *follow* the age of cessation of growth, a point which was first stressed by Gadgil and Bossert (1970). The prediction seems to be fairly well borne out by the evidence; although there are many species in which growth is prolonged past sexual maturity, there are only a few examples where growth ceases much before maturity. These will be discussed in section 5.4.2. As pointed out by Gadgil and Bossert, the present type of model would lead one to expect the times of sexual maturity and cessation of growth to coincide most closely in species where growth is rapid and then stops abruptly, as in holometabolous insects, since the maximum of $\tilde{l}(x)$ will then be close to the point at which $G(x)$ levels off (Figure 5.2). In species where slow growth continues for a long time, on the other hand, we would expect sexual maturity to be considerably in advance of the cessation of growth, as in cold-blooded vertebrates.

The reproductive effort model.   Most of the work in optimum life-history theory has been devoted to the 'reproductive effort model', originally suggested in verbal form by Williams (1966a, b), and first studied quantitatively by Gadgil and Bossert (1970). Later investigations include those of Fagen (1972), Schaffer (1974a, b), Schaffer and Gadgil (1975), Charlesworth and León (1976), Ricklefs (1977) Schaffer and Rosenzweig (1977) and Michod (1979). The basic idea is that, at any age x an individual can allocate a fraction $E_x (0 \leq E_x \leq 1)$ of its available energy resources to reproduction, and the remainder to maintenance (survival) and growth. E_x is called the *reproductive effort* at age x. Using the model developed on p. 234, we can represent the fecundity per unit weight at age x as a strictly increasing function of E_x, $\tilde{m}_x(E_x)$; the product of growth and survival at age x is a strictly decreasing function of E_x, $\tilde{P}_x(E_x)$ (Schaffer, 1974a). (The subscript x attached to $\tilde{m}$ and $\tilde{P}$ is intended to emphasise that, in general, the nature of the functional dependence of these variables on reproductive effort will vary with age.)

The optimisation problem in this case is thus to determine the vectors of values of E_x which yield maxima in r, in the density-independent case, or $\hat{N}$, in the density-dependent case. Since fecundity per unit weight is assumed to be an increasing function of effort, it is convenient in practice to use it directly as an independent variable, $\tilde{m}_x$, and to write $\tilde{P}_x$ as a function of $\tilde{m}_x$, $\tilde{P}_x(\tilde{m}_x)$. When reproductive effort at age x is unity, $\tilde{m}_x$ takes a maximal value M_x, such that $\tilde{P}_x(M_x) = 0$ When reproductive effort at age x is zero, $\tilde{P}_x$ takes its maximal value $\tilde{P}_x(0)$. Various forms of functional relationships between P_x and $\tilde{m}_x$ can be imagined (Figure 5.3); most of the results discussed below are based on the simplest (concave and convex) forms. Finally, it is important to note that parental care may form an important component of reproductive effort in some species; since we imagine that the survival of the young is an increasing function of the amount of parental care, this factor can be incorporated into the models by counting offspring at the age at which parental care ceases, so that $\tilde{m}_x$ includes this survival probability.

The problem of solving for the optimal life-history with this model is a formidable one, and no general analytic solutions have been found. Schaffer (1974*a*) and Schaffer and Rosenzweig (1977) have studied some special cases, and numerical solutions have been obtained by Gadgil and Bossert (1970) and Fagen (1972). Some insights into the properties of optimal life-histories can, however, be gained without knowing the full solutions, and these will be discussed below.

We first examine the conditions for r to be at a maximum with this model. Consider those age-classes for which reproductive effort is

Figure 5.3. Some possible functional relationships between $\tilde{P}_x$ and $\tilde{m}_x$: concave (full line), convex (dashed line) and concavo-convex (dotted line). $\tilde{P}_x$ and $\tilde{m}_x$ are in arbitrary units.

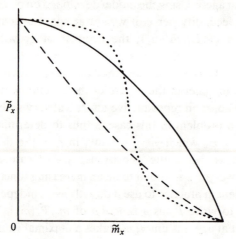

intermediate ($0 < \tilde{m}_x < M_x$). In order for r to be at a local maximum under such conditions, we need $\partial r/\partial \tilde{m}_x = 0$ for all such age-classes. Using the derivatives of r given in equations (5.4) and (5.11), we have

$$T\frac{\partial r}{\partial \tilde{m}_x} = e^{-rx}\tilde{l}(x)\left[1 + \frac{\partial \tilde{P}}{\partial \tilde{m}_x}e^{-r}\tilde{v}(x+1)\right] \tag{5.40}$$

The quantity $\tilde{v}(x+1)$ is Schaffer's *modified reproductive value* for age $x+1$, which is analogous to Fisher's reproductive value (equations (1.54)), but takes growth as well as survival into account. We have

$$\tilde{v}(x) = \frac{e^{rx}}{\tilde{l}(x)}\sum_{y=x}^{d}e^{-ry}\tilde{l}(y)\tilde{m}(y) \tag{5.41}$$

A necessary condition for a maximum in r is, therefore, that for age-classes with intermediate levels of reproductive effort, we must have†

$$1 + \frac{\partial \tilde{P}_x}{\partial \tilde{m}_x}e^{-r}\tilde{v}(x+1) = 0 \tag{5.42}$$

The second-order derivatives of r for all pairs of age-classes x and y such that $\partial r/\partial \tilde{m}_x = \partial r/\partial \tilde{m}_y = 0$ are given by

$$T\frac{\partial^2 r}{\partial \tilde{m}_x^2} = \frac{\partial^2 \tilde{P}_x}{\partial \tilde{m}_x^2}e^{-r(x+1)}\tilde{l}(x)\tilde{v}(x+1) \tag{5.43a}$$

$$\frac{\partial^2 r}{\partial \tilde{m}_x \partial \tilde{m}_y} = 0 \tag{5.43b}$$

Using the standard analytical conditions for a maximum of a function of several variables, the necessary and sufficient conditions for r to be maximised, with respect to perturbations in the age-classes with intermediate effort, are thus given by equation (5.42)

---

† An alternative procedure is used by Schaffer (1974*a*, *et seq.*). He gives the first age-class the index 0 instead of 1, and works with $B_x = \tilde{P}(0)\tilde{m}_x$ as the independent variable, instead of $\tilde{m}_x$. On this scale, $\tilde{v}_x$ is defined as $e^{-r}\tilde{P}(0)$ times the value of $\tilde{v}_x$ on the present scale (see the footnote to p. 37. If r is differentiated with respect to B_x, we obtain the equivalent of equation (5.42) as $1 + (\partial \tilde{P}_x/\partial B_x)\tilde{v}_{x+1} = 0$. The two notations yield identical conclusions.

and the following inequality

$$\frac{\partial^2 \tilde{P}_x}{\partial \tilde{m}_x^2} < 0 \tag{5.44}$$

The necessary and sufficient conditions for r to be at a maximum with respect to perturbations at age-classes with $\tilde{m}_x = 0$ or $\tilde{m}_x = M_x$ (i.e. boundary values of $\tilde{m}_x$) are

$$\frac{\partial r}{\partial \tilde{m}_x} < 0 \quad (\tilde{m}_x = 0) \tag{5.45a}$$

$$\frac{\partial r}{\partial \tilde{m}_x} > 0 \quad (\tilde{m}_x = M_x) \tag{5.45b}$$

In general, of course, perturbations to the life-history need not be confined to sets of age-classes within one of these three divisions. For small perturbations, it is easy to see that, if the above conditions are met, any change in r will be dominated by terms in which $\tilde{m}_x$ is on a boundary, since the total differential of r, dr, is zero with respect to perturbations in the other age-classes. Conditions (5.42), (5.44) and (5.45) are thus necessary and sufficient for r to be at a local maximum.

Similar conditions can be derived for the case of a density-dependent population, using the method of p. 233. We obtain the equivalents of expressions (5.40), (5.42), (5.43) and (5.44), for example, as

$$\frac{\partial w}{\partial \tilde{m}_x} = \tilde{l}(x)\left[1 + \frac{\partial \tilde{P}_x}{\partial \tilde{m}_x} \hat{v}(x+1)\right] = 0 \tag{5.46}$$

$$\frac{\partial^2 w}{\partial \tilde{m}_x^2} = \frac{\partial^2 \tilde{P}_x}{\partial \tilde{m}_x^2} \tilde{l}(x)\hat{v}(x+1) < 0 \tag{5.47}$$

In carrying out these differentiations, it is important to note that the demographic parameters in these equations are held constant at the values for population density equal to the carrying-capacity given by equation (5.31a) (Charlesworth and León, 1976).

Properties of optimum life-histories with the reproductive effort model: semelparity. Some useful results can be obtained merely by inspection of the above formulae. We may note first of all that a necessary condition for inequality (5.44) to be satisfied is that $\tilde{P}_x$ be a concave function of $\tilde{m}_x$ for at least part of its range (cf. Gadgil and Bossert, 1970; Schaffer, 1974a). If $\tilde{P}_x$ is a convex function of $\tilde{m}_x$, $\partial^2 \tilde{P}_x / \partial \tilde{m}_x^2$ is necessarily positive throughout its range, so that there can be no age-classes with intermediate levels of reproductive effort; the life-history in such a case must be semelparous, with a number of non-reproductive age-classes terminating in an age-class with maximal reproductive effort. This, as we have discussed earlier, is characteristic of a number of groups, and it is tempting to speculate that such semelparous life-histories are associated with convex relationships between $\tilde{P}_x$ and $\tilde{m}_x$. There is, unfortunately, little direct evidence on this point. Gadgil and Bossert (1970) suggested that semelparous life-histories characteristic of Pacific salmon involve convexity; these species undertake long migrations from the ocean to rivers in order to breed, so that much energy has to be expended, and exposure to considerable risks undertaken, in order for the salmon to breed at all. This means that the curve relating $\tilde{m}_x$ to E_x rises very slowly at first, and then increases rapidly, whereas $\tilde{P}_x$ is reduced considerably by a small increase in E_x. This is likely to generate a convex relationship between $\tilde{P}_x$ and $\tilde{m}_x$. Similarly, Schaffer and Gadgil (1975) proposed that the contrast between the semelparous agaves and the phenotypically similar but iteroparous yuccas may be traceable to greater competition for pollinators in the former, the insects preferring large inflorescences which require large stalks. Evidence for such preferences is presented by Schaffer and Schaffer (1977, 1979). Since a large expenditure of energy is probably needed to develop an inflorescence stalk, which is required for even a low level of reproduction, a convex relationship between $\tilde{P}_x$ and $\tilde{m}_x$ may be generated. This is particularly likely in a competitive situation, since plants with relatively small stalks may have difficulty in attracting pollinators if others with larger stalks are present, despite a significant expenditure of effort.

It should be noted that it is possible for there to be several different semelparous life-histories which are local maxima in r or $\hat{N}$, for the same set of assumed relationships between $\tilde{P}_x$ and $\tilde{m}_x$. Schaffer and Rosenzweig (1977) give some examples of this. Each of these life-histories is, by definition, immune to invasion by mutants which cause a small increase in fecundity at some age other than that at which reproduction is currently taking place, but may be susceptible to invasion by a mutation which creates a life-history that has reproduction shifted to a value of b corresponding to greater fitness. If genetic variation in the timing of reproduction is possible, we would therefore expect b to be adjusted to the value which gives a global maximum in fitness (cf. section 5.3.1).

Properties of optimum life-histories with the reproductive effort model: iteroparous life-histories. For studying iteroparous life-histories in which there is a number of age-classes with intermediate reproductive effort, it is useful to restrict attention to the case when the $\tilde{P}_x$ have the same, concave dependence on the $\tilde{m}_x$ for each age-class with intermediate reproductive effort. $\tilde{P}_x(\tilde{m}_x)$ can therefore be written simply as $\tilde{P}(\tilde{m}_x)$. The first result that we can derive in this case is that there must be an inverse relationship between $\tilde{m}_x$ and $\tilde{v}(x+1)$ for all ages with intermediate reproductive effort. By the hypothesis of concavity, $|\partial\tilde{P}/\partial\tilde{m}_x|$ is an increasing function of $\tilde{m}_x$. Equation (5.42) thus implies that, if we compare two ages x and y, $\tilde{m}_x > \tilde{m}_y$ if and only if $\tilde{v}(x+1) < \tilde{v}(y+1)$. A relationship of this sort, but between Fisherian reproductive value and reproductive effort, was first proposed by Williams (1966b) on intuitive grounds. He suggested that what he called *residual reproductive value* at age x, defined as the component of reproductive value attributable to future reproduction, should be inversely related to reproductive effort at age x. From equation (5.41) it is evident that we have

$$\tilde{v}(x) = \tilde{m}(x) + e^{-r}\tilde{P}(x)\tilde{v}(x+1) \qquad (5.48)$$

Residual reproductive effort in this formulation is thus equal to $e^{-r}\tilde{P}(x)\tilde{v}(x+1)$, so that Williams' suggestion and the use of it by Pianka and Parker (1975) are not quantitatively correct.

Another question which can be answered without knowing the full details of the optimal life-history solutions concerns the relations between reproductive effort and age. It was suggested by Williams (1966a, p. 182) and by Gadgil and Bossert (1970) that reproductive effort should increase with age, since the cost of present reproduction to future fitness depends on the number of future reproductive age-classes, which clearly decreases with increasing age. Gadgil and Bossert constructed some computer examples of optimal life-histories which always showed an increase in E_x with x for the ages with intermediate effort. Some counter-examples were, however, produced by Fagen (1972). An algebraic treatment was given by Charlesworth and León (1976), whose results are outlined below. In brief, they found that reproductive effort may increase with age, but need not necessarily do so.

We shall retain the assumption that $\tilde{P}(\tilde{m}_x)$ is the same concave function of $\tilde{m}_x$ for each age-class with intermediate effort, so that the only constraints placed on the relationship between effort and age are those due to the nature of the optimal life-history solution. We know that $\tilde{m}_{x-1} < \tilde{m}_x$ if and only if $\tilde{v}(x+1) < \tilde{v}(x)$ for the optimal life-history. Expressing $\tilde{v}(x)$ in terms of $\tilde{v}(x+1)$ by means of equation (5.48), the condition for $\tilde{m}_{x-1} < \tilde{m}_x$ in the density-independent case is thus

$$\tilde{v}(x) = \tilde{m}_x + e^{-r}\tilde{P}(m_x)\tilde{v}(x+1) > \tilde{v}(x+1) \tag{5.49}$$

The variables r and $\tilde{v}$ in this equation are, of course, understood to take values which correspond to the optimal life-history defined by equations (5.42) and (5.45). If r and $\tilde{v}(x+1)$ are held constant at these values, we obtain

$$\frac{\partial \tilde{v}(x)}{\partial \tilde{m}_x} = 1 + \frac{\partial \tilde{P}}{\partial \tilde{m}_x} e^{-r}\tilde{v}(x+1) \tag{5.50}$$

From equation (5.42), $\partial \tilde{v}(x)/\partial \tilde{m}_x$ is zero when $\tilde{m}_x$ is equal to its value for the optimal life-history; from the concavity of $\tilde{P}(\tilde{m}_x)$, it must be positive for smaller values of $\tilde{m}_x$, and negative for larger values. $\tilde{v}(x)$ is thus at a *global* maximum with respect to changes in $\tilde{m}_x$ alone, when the latter is equal to its value for the optimal life-history.[†] From equation (5.49), a sufficient condition for $\tilde{v}(x) > \tilde{v}(x+1)$ is thus

$$e^{-r}\tilde{P}(0) \geqq 1 \tag{5.51}$$

[†] It is interesting to note that it can be shown that $\tilde{v}(x)$ is at a *local* maximum at the optimal life-history with respect to all perturbations to the life-history at age-classes with intermediate reproductive effort (see p. 266). This does not require r to be held constant. In this sense, therefore, maximisation of r corresponds to maximisation of reproductive value, in the reproductive effort model (Schaffer, 1974a).

This relation provides a sufficient condition for reproductive effort to increase with age throughout the reproductive period. It is obviously easiest to satisfy if r for the optimal life-history is not too high and if growth and survival rates are high for low levels of reproductive effort. This is in accordance with common sense, since it is under these conditions that postponement of reproduction is least disadvantageous, as we have already seen. If there is no growth of reproductively mature individuals, $\tilde{P}(0)$ is the maximum survival probability, which must be less than one. Condition (5.51) may easily be violated in such a case, which leaves the relationship between reproductive effort and age undetermined. A *necessary* condition for reproductive effort to increase with age can be derived as follows. From equation (5.49), $\tilde{v}(x) > \tilde{v}(x+1)$ implies that

$$\frac{e^r \tilde{m}_x}{\tilde{v}(x+1)} + \tilde{P}(\tilde{m}_x) = \frac{e^r \tilde{v}(x)}{\tilde{v}(x+1)} > e^r \qquad (5.52)$$

Let K be the maximum value of $|\partial \tilde{P}/\partial \tilde{m}_x|$, attained when $\tilde{m}_x = M$. Equation (5.42) implies that the optimal life-history must satisfy the condition

$$K e^{-r} \tilde{v}(x+1) \geqq 1$$

A necessary condition for inequality (5.52) to be satisfied is thus

$$K\tilde{m}_x + \tilde{P}(\tilde{m}_x) > e^r$$

Taking the derivative with respect to $\tilde{m}_x$, it is easily seen that the left-hand side of this inequality increases with $\tilde{m}_x$, so that a necessary condition for $\tilde{v}(x+1) < \tilde{v}(x)$ $(\tilde{m}_{x-1} < \tilde{m}_x)$ is

$$KM > e^r \qquad (5.53)$$

If r for the optimal life-history is high, survival and growth do not fall off too much with increasing reproductive effort, and the maximal fecundity per unit weight is low, then this inequality may be violated, and reproductive effort will decrease with age or remain constant. These are circumstances in which intuition would lead us to expect selection to place a premium on reproduction early in life. It is thus theoretically possible for reproductive effort not to increase with age. As shown by Charlesworth and León (1976), satisfaction of condition (5.51) automatically implies satisfaction of condition (5.53), but the converse is not true. There may, therefore, be non-monotonic relationships between age and reproductive effort under some conditions.

A necessary condition for $\tilde{m}_x < \tilde{m}_{x-1}$ can be derived on similar lines to (5.53). Let k be the minimum value of $|\partial \tilde{P}/\partial \tilde{m}_x|$, attained when $\tilde{m}_x = 0$. The condition is

$$kM < e^r \qquad (5.54)$$

This inequality is likely to be violated under opposite conditions to those which favour violation of inequality (5.53); violation of (5.54) provides a sufficient condition for reproductive effort to increase with age, or remain constant.

It is of some interest to determine the conditions under which reproductive effort remains constant for each adult age-class in an optimal life-history, particularly as several analyses have been carried out on the assumption that this is so (Schaffer, 1974*a*, *b*; Ricklefs, 1977). In such a case, we can write $\tilde{m}_x = \tilde{m}$ and $\tilde{P}(\tilde{m}_x) = \tilde{P}$ for each adult age-class in the optimal life-history. If we assume that there is no upper limit to the age to which individuals can survive and reproduce (equation (5.21*a*) and substitute the above values for m_x and $\tilde{P}(\tilde{m}_x)$ into equation (5.41) we find that

$$\tilde{v}(x) = e^{rx}/\tilde{l}(x) \quad (x < b) \qquad (5.55a)$$

$$\tilde{v}(x) = \tilde{m}/(1 - e^{-r}\tilde{P}) \quad (x \geqq b) \qquad (5.55b)$$

Equation (5.42) thus gives us the condition on the adult age-classes

$$\left| \frac{\partial \tilde{P}}{\partial \tilde{m}} \right| = \frac{e^r - \tilde{P}}{\tilde{m}} \qquad (5.56a)$$

Using equation (5.39), this can be expressed as

$$\left| \frac{\partial \tilde{P}}{\partial \tilde{m}} \right| = e^{-r(b-1)}\tilde{l}(b) \qquad (5.56b)$$

Provided that $k < e^{-r(b-1)}\tilde{l}(b) < K$, this condition can be satisfied, and an optimal life-history with constant adult reproductive effort will exist, as far as perturbations in the adult age-classes are concerned. If $k = 0$ and $K = \infty$, for example, the condition is automatically satisfied. When $b = 1$, the condition is quite independent of the value of r.

We must also consider perturbations to the juvenile age-classes. In order for conditions $(5.45a)$ to be satisfied, we must have

$$\left| \frac{\partial P_x}{\partial \tilde{m}_x} \right| > e^{-rx} \tilde{l}(x+1) \quad (1 \leqq x < b, \, \tilde{m}_x = 0) \tag{5.57}$$

Joint satisfaction of conditions (5.56) and (5.57) requires that $\tilde{P}_x$ be more sensitive to increases in $\tilde{m}_x$ for $x = b - 1$ than for the later age-classes. In organisms in which $r \geqq 0$ and growth is completed by age 1, so that $\tilde{l}(x)$ may be taken equal to $l(x)$, condition (5.57) is satisfied for $1 < x \leqq b$ whenever condition (5.57) holds for $x = 1$, provided that $\tilde{P}_x$ for $x > 1$ is at least as sensitive to increases in $\tilde{m}_x$ as for $x = 1$. In other cases, greater sensitivity of $\tilde{P}_x$ to increased fecundity may be required for later juvenile age-classes. In conclusion it is not difficult, in principle, to find cases in which an optimal life-history exists which has constant reproductive effort among the adult age-classes, and hence constant adult survival and fecundity.

The above results on the relations between reproductive effort and age can easily be extended to density-dependent populations, using equations (5.46) and (5.47). The consequences of density dependence vary according to the model assumed. If density dependence of an age-independent component of $\tilde{m}_x$ is assumed (equation (5.36)), we can write $\tilde{m}_x = \tilde{m}_x^{(I)} \tilde{m}^{(D)}(N)$. If the density-dependent component is unrelated to reproductive effort, $\tilde{P}_x$ is a function of $\tilde{m}_x^{(I)}$ alone. Substituting into equation (5.46), we obtain the condition for an optimal life-history

$$\tilde{m}^{(D)}(\hat{N}) \left[1 + \frac{\partial \tilde{P}_x}{\partial m_x^{(I)}} \tilde{v}(x+1) \right] = 0 \tag{5.58}$$

The term $\tilde{m}^{(D)}(N)$ can thus be factored out of the equations for an optimal life-history. Conditions similar to those given by expressions (5.51)–(5.54) can thus be obtained, except that r is set equal to zero. This makes it easier to satisfy the conditions for reproductive effort to increase with age. A similar result is obtained if a dependence of pre-reproductive survival or growth on density is assumed.

If, however, density regulation is mediated through an age-independent component of growth or survival, $\tilde{P}_D(N)$, so that $\tilde{P}_x = \tilde{P}_x^{(I)}(\tilde{m}_x)\tilde{P}_D(N)$ (cf. equation (5.37)), equation (5.58) becomes

$$1 + \frac{\partial P_x^{(I)}}{\partial \tilde{m}_x}\tilde{P}_D(\hat{N})\tilde{v}(x+1) = 0 \qquad (5.59)$$

$P_D(\hat{N})$ thus plays the same role as e^{-r} in the density-independent case, so that there is no reason to expect it to be any easier for reproductive effort to increase with age. These conclusions about the effects of different models of density dependence are very similar to those reached in connection with the problem of the optimum age of reproductive maturity (pp. 235–6).

Other consequences of the reproductive effort model.   It should be noted that the age of reproductive maturity *b* is itself a component of the optimal life-history solution in the reproductive effort model, and in other types of optimal life-history models. Numerical examples are given by Gadgil and Bossert (1970). Considerations such as those discussed on pp. 233–7 play a role in determining the age of maturity. Another important factor is likely to be dependence of the functional relationship between $\tilde{P}_x$ and $\tilde{m}_x$ on age (p. 237). Several authors (e.g. Lack, 1954, Chapter 6) have suggested that reproduction may impose a greater strain on very young individuals, or that a greater level of reproductive effort might be needed for a given level of reproductive success by young, inexperienced individuals. Both these factors will tend to cause $|\partial \tilde{P}_x/\partial \tilde{m}_x|$ to decrease with age among the younger age-classes, and hence for postponed reproduction to be favoured. An extreme example of this kind of situation is provided by *Drosophila*, where it is known that the larva must attain a certain minimum weight in order to pupate successfully (Bakker, 1959). Any attempt to advance the time of maturation, beyond a point at which larvae cannot reach the critical weight, must lead to lethality.

It has been suggested (Gadgil and Bossert, 1970; Schaffer 1974*a*) that the level of reproductive effort at each age, in an optimum life-history, will depend on the level of extrinsic causes of mortality, such as predation or density-dependent factors. Thus, if adult mortality is high, the optimum life-history should have a higher level of reproductive effort at each age than if adult mortality is low. Conversely, a low level of reproductive success or juvenile survival will tend to favour a low level of reproductive effort. Analytical results concerning this question are hard to obtain because of the difficulty of solving for the optimal life-history. Schaffer's treatment assumes that the optimal life-history is such that reproductive effort for the adults is independent of age, and that reproductive maturity occurs at age $b = 1$. Equation (5.5b) for the optimal life-history in a density-independent environment thus becomes

$$\left| \frac{\partial \tilde{P}}{\partial \tilde{m}} \right| = \tilde{l}(1) \tag{5.60a}$$

Assume that $\tilde{P}(\tilde{m})$ is changed by an increase in some extrinsic source of mortality to $P'\tilde{P}(\tilde{m})$, where P' is less than one and independent of reproductive effort. The equivalent of equation (5.60a) for the optimal life-history under the new regime of adult mortality is

$$\tilde{P}' \left| \frac{\partial \tilde{P}}{\partial \tilde{m}} \right| = \tilde{l}(1) \tag{5.60b}$$

If $\tilde{P}$ is a concave function of $\tilde{m}$, satisfaction of this relation requires that the new optimal life history have a higher value of $\tilde{m}$ than the old, in accordance with the idea that higher adult mortality should lead to higher reproductive effort. On the other hand, if $\tilde{l}(1)$ or $\tilde{m}$ are changed to new, lower values in a similar fashion, it is easily seen that the new optimal life-history should have lower reproductive effort.

We have seen above that reproductive effort is not always independent of age in the optimal life-history, so that these results must be treated with some caution as a generalisation. Some progress in analysing the problem by a more general approach has recently been made by Michod (1979). Using partial differentiation, he was able to obtain expressions for the changes in the $\tilde{m}_x$ values for the optimal life-history which are induced by a small change in the probability of survival at an arbitrary age y. More explicitly, if $\tilde{P}_y(\tilde{m}_y)$ is changed to a new value $P'_y\tilde{P}_y(\tilde{m}_y)$, as the result of a change in some extrinsic cause of mortality, the new solution for the optimal life-history differs from the old in a way which is calculable from Michod's formulae, provided that P'_y is close to unity.

The exact consequences of a change in survival at one age-class depend on whether or not there is density dependence and, if there is, on its exact form. In density-independent populations, an increase in mortality at age y will generate a new optimal life-history which has greater reproductive effort at each age up to and including y, but lower reproductive effort following age y. Thus, an increase in mortality among the juveniles will lead to a decrease in reproductive effort among the adults in this case, and a new source of mortality late in life will lead to an increase in reproductive effort at earlier age-classes. With density dependence, it is still true that an increase in mortality at age y leads to an increase in reproductive effort at all reproductive ages $x \leqq y$; the consequences for ages $x > y$ depend, however, on the specific model of density dependence assumed. If density dependence operates in such a way that $\tilde{m}_x$ and $\tilde{P}_x$ are insensitive to density for $x > u$ (where u is some fixed age), then there is *no* effect on reproductive effort at ages later than u. Thus, a change in mortality among the juvenile age-classes when there is density-dependent juvenile survival has no effect on the reproductive effort of the adult age-classes, in contrast to a reduction of reproductive effort in a density-independent population. The effects of an increase in mortality in several age-classes at once can be complex. It is still true, however, that all reproductive ages prior to any ages at which mortality is increased will increase their reproductive effort. In a density-independent population, there will also be a reduction in reproductive effort at each age greater than the last age to be affected by a mortality change. For other ages, the nature of the changes is difficult to determine. In a population regulated by density-dependent juvenile survival, we would expect an increase in reproductive effort at each reproductive age up to and including the last reproductive age-class at which there is an increase in mortality, and no change thereafter. Finally, the effects of an increase in mortality applied equally to each age-class also depend on the mechanism of density-dependence. In a density-independent population, such a change is exactly compensated for by a reduction in r (p. 35), so that there is no change in the characteristic equation (5.30a) for the optimal life-history, and hence no change in the distribution of reproductive effort (Gadgil and Bossert, 1970). This is also true in a population regulated by an age-independent, density-dependent component of survival (equation (5.37), but in a population regulated by density-dependent juvenile survival, an increase in reproductive effort of the adults is to be expected as a result of this type of mortality change.

Evidence for the reproductive effort model. Empirical evidence concerning the relevance of the reproductive effort model as a factor in life-history evolution comes from two different sources. The first involves direct evidence for the type of relationships between survival, growth, fertility and reproductive effort which is assumed by the model. This comes from several different organisms. Maynard Smith (1958) showed that laying eggs caused a reduction in the probability of subsequent survival of female *Drosophila subobscura*; similarly, Gowen and Johnson (1946) and Rose (1979) observed a negative correlation between longevity and daily rate of egg production in *D. melanogaster*. Lawlor (1976) showed that the amount of growth at moulting of female woodlice (*Armadillidium vulgare*) is lower in reproductive than in comparable non-reproductive females. The estimated energy available for growth and reproduction in reproductives is similar to that available for growth alone in non-reproductives. Snell and King (1977) obtained data on a species of rotifer that demonstrated a negative correlation between fecundity at a given age and the probability of subsequent survival. In shrubs and trees, it is frequently observed that fruit set fluctuates between low and high values in alternate years, suggesting that reproduction imposes a strain from which the plant needs to recover (Harper and White, 1974). Other examples are reviewed by Stearns (1976). Detailed information on the form of the functional relationships postulated in Figure 5.3 is almost completely lacking, however, as is information over how long the adverse effects of an increase in reproduction at a given age are likely to persist.

The other source of evidence comes from demographic studies of natural populations, which might be expected to provide information on any relationships between reproductive effort and age, or reproductive value and reproductive effort, of the sort discussed above (pp. 242–7). The results here are even more scanty than those from the direct studies. One of the chief difficulties is that of actually measuring reproductive effort, particularly in species where parental care must be regarded as a component. The most that can be done, generally speaking, is to use the fraction of body weight taken up by the reproductive organs, or the corresponding fraction of total calories, as a crude estimate of reproductive effort. We have seen that reproductive effort is particularly likely to increase with age in groups where adults continue to grow and have high survival probabilities, and where the intrinsic rate of increase is low (pp. 244–5). There are some cases of apparent increase of reproductive effort with age in species of fish and reptiles which probably meet these criteria (Gerking, 1959; Hodder, 1963; Tinkle and Hadley, 1975). There are also cases in which this does not seem to be the case (Tinkle and Hadley, 1975). Whether these differences are related to differences in ecological factors which might affect the mathematical conditions for increase or decrease of reproductive effort is unknown. A stronger test of the theory might be to use the inverse relationship between modified reproductive value, $\tilde{v}(x)$, and reproductive effort for adult age-classes, which is expected when $\tilde{P}_x$ is the same function of $\tilde{m}_x$ for each adult age-class (p. 242). The analysis of lizard data by Pianka and Parker (1975) shows that this relationship holds in some cases, but not in others. Unfortunately, the data are usually available only in a form which enables Fisherian reproductive value to be calculated; as we have seen, this is only of direct relevance when there is no growth of adults. This type of data has not, as yet, provided any conclusive tests of the reproductive effort model. In many cases, reproductive value decreases with advancing age somewhat after reproductive maturity (Table 1.4), even when fecundity is constant or declines with age. Unless there is some compensatory increase in juvenile survival due to increased parental care, which can hardly apply to *Drosophila*, these observations are not consistent with the reproductive effort model. They do fit with the senescence models discussed in sections 5.2.2, which of course do not assume that the life-history is optimal. If we allow changes in the form of the dependence of $\tilde{P}_x$ on $\tilde{m}_x$ with age, almost any relationships between reproductive effort, reproductive value and age can be generated.

This would amount to an admission that other factors beside reproductive effort are important in controlling life-history evolution.

5.4   Conclusions
In this section, we shall survey the main conclusions which can be derived from the results described above, and try to relate them to some of the observations which have been made on life-history phenomena. Some special aspects of life-history evolution will also be reviewed briefly.

5.4.1   *The evolution of senescence*
The results discussed in the first part of this chapter (section 5.2) suggest that the idea that selection acts more strongly on genes which affect survival in pre-reproductive life or early in reproductive life, compared with genes that act later, provides an adequate explanation of the general features of senescence. The predictions of this theory concerning the effects of differences in demographic conditions on the strength of selection for senescence seem to be verified, at least in general outline (section 5.2.2). It should be clear from the discussion that a variety of physiological factors are likely to be involved in ageing processes, since senescence is imagined to evolve as the result of the age-specific effects of genetic variation at a large number of loci (Williams, 1957). It is, accordingly, unlikely that a single process can be the sole cause of senescence. The evolutionary models also imply, as we have seen, that senescence is almost certain to evolve whenever there is separation of soma and germ-plasm. This is true even when the initial life-history is non-senescent, in the sense of having age-specific survival probabilities which do not decrease with age.

Kirkwood (1977) and Kirkwood and Holliday (1979) have criticised this view of the evolutionary origin of ageing, on the grounds that the existence of the age-specific gene effects demanded by the models implies that genes or their products must be capable of detecting the passage of time during the life-history. According to Kirkwood and Holliday, this in turn implies that ageing must already exist, in the sense of biochemical distinctions between different ages. This argument seems, however, to equate senescence with any kind of physiological differences associated with age, whereas the concept is usually restricted to progressive deterioration with age. Even after the completion of development and the attainment of reproductive maturity, non-senescent life-history changes such as growth in size seem likely to have occurred in primitive multicellular organisms, and could have provided the necessary clocks to which age-specific gene effects might respond. Furthermore, the process of development up to the age of reproductive maturity implies the existence of a changing biochemical environment in part of the life-history, by means of which gene activity can be regulated. It is easy, for example, to imagine genes which are turned on before the end of development, but whose products decay over an interval of time that extends into the reproductive period of life.

Kirkwood and Holliday have also put forward an interesting hypothesis which places Orgel's (1963, 1973) error-catastrophe theory of ageing in an evolutionary context. This theory is based on the fact that the machinery of protein synthesis and DNA replication is somewhat imperfect. Errors will thus be produced in the sequences of the proteins that are themselves involved in these processes. If the frequency of such errors is sufficiently high, positive feed-back of error frequency takes place in successive rounds of protein synthesis and DNA replication, so that there is a cumulative increase in the frequency of abnormal proteins produced by cells, leading eventually to cellular death. If the initial frequency of errors is sufficiently low, however, the frequency of abnormal proteins reaches a steady level (Orgel, 1973; Hoffman, 1974; Kirkwood and Holliday, 1975; Goel and Ycas, 1976), and no such error-catastrophe occurs. Kirkwood and Holliday suggests that there is an energetic cost to the proof-reading mechanisms needed to keep errors to a low, steady level. In a unicellular organism, a low error level is clearly essential for the continued existence of a clone of cells. In a species with separation of soma and germ-plasm, there could be a selective gain (due to energy-saving) in reducing the accuracy of proof-reading, at any rate after some period of life. Provided that this gain is sufficiently high, it may outweigh the disadvantages of a subsequent error-catastrophe and death. This follows from the considerations of selection as a function of age which were discussed earlier. There is no contradiction between this model and the evolutionary models described above; it can, indeed, be regarded as a special case of Williams' pleiotropy model (p. 217). Differences in life-span between species can, on this model, be interpreted in terms of different intensities of selection on the accuracy of proof-reading in different demographic environments. Any future experimental proofs of the importance of error-catastrophe in ageing can, therefore, easily be reconciled with the classical selective models for the evolution of ageing.

5.4.2 *Reproductive patterns and age*
The effects of different demographic conditions. We have seen that genes affecting fecundity will normally be selected for more strongly, the earlier the point in life at which they act. This differential in favour of early reproduction is most marked when there is a high level of mortality or a high rate of population growth so that $e^{-rx}l(x)$ falls off steeply with age (pp. 224–5). The effects of demographic conditions on selection with respect to timing of reproduction can be analysed in very much the same terms as their effects on gene frequencies in equilibrium populations under selection (section 3.4.1). We have seen that semelparity is most likely to be favoured over iteroparity when adult survival is low, relative to juvenile survival, and the population growth-rate is high (pp. 225–7), or when adult survival (as opposed to fecundity or juvenile survival) fluctuates in time (pp. 228–9). A high level of reproductive effort is likely to be favoured when adult survival is low, due to extrinsic factors such as predation, whereas high juvenile mortality favours lower reproductive effort (pp. 247–50). Reproductive effort is most likely to increase with age when population growth is low and adult survival high, and when individuals continue to grow in size or reproductive efficiency during adult life (pp. 242–7). Postponement of reproductive maturity is most likely when immature individuals have low mortality and high growth-rates compared with adults, and when fecundity and population growth are low.

There has been much discussion in the literature concerning the existence of correlations of this sort between the demographic conditions of species and the types of life-history that they exhibit

(Lewontin, 1965; MacArthur and Wilson, 1967, Chapters 4 and 7; Gadgil and Bossert, 1970; Pianka, 1970; Southwood *et al.*, 1974, Schaffer and Gadgil, 1975). This has led to stress being laid on the contrast between '*r*- selection and *K*-selection' (MacArthur and Wilson, 1967). *r*-selection is said to operate on species which are relatively free of density-dependent regulation, but may have high density-independent mortality, and whose populations therefore spend much of their time growing, but crash when conditions become unfavourable. As we have seen, selection favours a high intrinsic rate of increase in such circumstances. This kind of situation is exemplified by species which occupy temporary habitats, such as weeds, or by species which have seasonal periods of high population growth, such as temperate-zone insects. *K*-selection, on the other hand, is said to occur when a species lives in a density-dependent environment with a relatively constant population size. Under these conditions, we have seen that natural selection often tends to maximise the carrying-capacity (K, in the notation used by MacArthur and Wilson). In contrast to *r*-selection, this is said to lead to slow growth, late maturity, low reproductive effort, and iteroparity rather than semelparity. Examples of *K*-selected species include forest trees and large mammals and birds.

While the distinction between these modes of selection has considerable value, it should be remembered that the original formulations of *r*-selection and *K*-selection were in terms of models that largely neglected the factor of age-structure, and were based on the logistic growth equation described by equation (1.71) (MacArthur, 1962; MacArthur and Wilson, 1967). It is not very surprising that some important aspects of the impact of demographic factors on the outcome of selection are missing from the *r*-selection versus *K*-selection contrast. For instance, we have seen in several contexts that the effects of density regulation by means of a mortality factor applied equally to each age-class are similar to the effects of a high rate of population growth (pp. 35, 236, 247). If we were to compare two populations which differed only in that one was held in check by this form of density dependence, and the other was density independent, no differences in the outcome of selection would be expected. On the other hand, population regulation by means of density-dependent juvenile survival or fecundity reduces the intensity of selection for early breeding, compared with a density-independent population (p. 235). As in the cases discussed in section 3.4.1, differences in age-structure are the important factors. We have also seen that temporal fluctuations in survival rates and fecundities may have consequences for the form of the life-history which are not included within the compass of r-selection and K-selection (pp. 228–9). It would therefore seem wisest to investigate the effects of demographic factors on the outcome of life-history evolution by means of models which take account of details of the age-specificity of these factors, rather than by appealing to the blanket concept of r-selection and K-selection. Similar views have been expressed by Wilbur, Tinkle and Collins (1974) and Mertz (1975).

Empirical tests of life-history theory. The fact that the general predictions of life-history theory seem to be borne out by species comparisons hardly constitutes a proof of the theory. Critical tests are difficult to construct. For example, we have seen that it is hard to say whether or not detailed demographic data on natural populations are consistent with the reproductive effort model (pp. 250–2). Stearns (1976, 1977) has come to a similar conclusion, on the basis of an extensive review of the literature. As pointed out by Stearns, there is often difficulty in discriminating between alternative explanations of the same piece of data, a problem which frequently occurs in evolutionary theory. There are numerous problems associated with attempts to test the theory by means of species comparisons. For example, multiple ecological differences, possibly unknown to the investigator, may make it hard to identify the factors which are truly responsible for observed life-history differences.

A better method of testing the theory is to make comparisons between populations of the same species which are subject to different demographic regimes. Few such studies have as yet been made. Solbrig and Simpson (1974) studied several apomictic strains of the dandelion, *Taraxacum officinale*, grown in standard culture conditions. One of these strains (A) predominated in a trampled area of land, and another (D) was commonest in a relatively protected area. It is likely that the former population experiences high density-independent mortality, due to disturbance, whereas the other is density-regulated, possibly because of density-dependent seedling mortality (Harper and White, 1974). It was found that strain A exhibited a higher reproductive effort than D, in terms of number of flowering heads and seeds per plant, but performed less well in growth-competition experiments with D. This agrees reasonably well with theoretical expectation. Unfortunately, no information on age-specific aspects of reproduction and survival are available for this case.

Information of this sort has been obtained for the annual meadow grass, *Poa annua*, by Law (1975, 1979) and Law, Bradshaw and Putwain (1977). Despite the species' name, the plants studied by these authors had a maximum life-span of about 18 months. Reproduction is initiated approximately 15 weeks after germination in winter, and continues through spring and summer of the first year. If the plant survives to the following year, a second period of reproduction in spring and summer precedes death. These plants are largely self-fertilising, so that reproduction is effectively clonal. Life-history data were collected on plants raised from seed in standard conditions, and derived from 29 different natural populations. These could be divided into two broad categories of roughly equal size, 'opportunist' and 'pasture'. The opportunist populations were subject to a good deal of density-independent mortality due to disturbance, and the pasture populations were relatively stable communities, subject to regulation by density-dependent seedling mortality. Differences between the two classes of population, as well as between populations within each class, were detected in a number of different life-history variables. Fecundity, as measured by number of inflorescences per plant, was much higher in the second than the first year of life for the pasture populations. Opportunist populations mostly had higher fecundity in the first year than the pasture populations, and a more even distribution of reproduction between the two years. They also reproduced significantly earlier within the first year than did the pasture populations, and the plants were smaller. The opportunist populations had higher mortality than the pasture populations; this was concentrated in the period following reproduction in the first year, which suggests that it is due to the adverse effect of reproductive effort on subsequent survival. There is no doubt that the differences observed in this study are genetic, since the plants were raised under the same conditions. The differences are consistent with theoretical expectations for the differences between populations subject to density-dependent juvenile mortality, and populations subject to high age-independent and density-independent mortality. The observation that fecundity increases with age in the density-dependent pasture populations is particularly interesting, in view of the predictions of the reproductive effort model discussed on pp. 243–7.

An example of postponed reproduction. A life-history that, at first sight, presents several problems for the theories discussed above is exhibited by island-breeding sea-birds which feed offshore, particularly tropical species. As remarked on by many authors (Lack (1954, Chapter 6; 1966, Chapter 16) Wynne-Edwards (1955; 1962, Chapter 5) Ashmole (1963) Goodman (1974)), these birds show a combination of traits associated with apparently low reproductive effort, including postponement of reproduction for several years, a clutch size of one, prolonged incubation and nestling periods, frequent abandonment of the nest by parents, and long intervals between successive clutches. The extreme example is provided by the royal and wandering albatrosses, *Diomedea epomophora* and *D. exulans*, which reach sexual maturity at about 10 years of age, have an incubation period of around 80 days, and a nestling period of about 250 days; since it takes over a year to rear a chick successfully, breeding occurs only once every two years (Lack, 1966; Goodman, 1974). These features have been interpreted by Wynne-Edwards (1955, 1962) as evidence that the life-history has evolved by group selection in favour of a low reproductive potential, in order to avoid exhaustion of the food supply and consequent extinction of the population. As mentioned on p. 205, there are strong theoretical objections to group selection, and we shall now consider some alternative interpretations of this form of life-history, which have been proposed by Ashmole (1963), Lack (1966) and Goodman (1974).

In the first place, the small clutch size of these species is understandable in terms of the difficulty they have in feeding their young;

since the birds are large in size and restricted to breeding on islands, many adults will be competing for food within foraging range, so that strong density-dependent mortality of nestlings would be expected. Evidence for such mortality is described by Ashmole (1963), Lack (1966) and Nelson (1969). Experimental addition of an extra chick to nests of the Laysan albatross, *D. immutabilis*, increases mortality due to starvation, to such an extent that there is a considerable net decrease in the number of chicks raised per pair (Rice and Kenyon, 1962). There is thus little difficulty in accounting for the small clutch size. The prolonged nestling period can be interpreted in terms of slow growth of the young, evolved in response to the limited food supply (Ashmole, 1963; Lack; 1966). The long incubation period of the eggs is more difficult to explain. As suggested by Lack, it may be impossible to prolong the nestling period except by a general slowing-down of growth, which also retards the development of the egg.

The main question of interest here is the reason for the prolonged period of sexual immaturity, despite the fact that growth is essentially completed before the birds can fly (Ashmole, 1963). There is good evidence for a very low annual mortality of full-grown birds (5–10%), presumably due to the virtual absence of predation (Goodman, 1974). The intensity of selection in favour of early breeding is therefore low. Because of the early completion of growth, any explanation based on the ideas discussed in section 5.3.2 has to be couched in terms of an increase in potential reproductive efficiency with age. As suggested by Ashmole (1963), it is probable that older birds are more efficient than younger ones in foraging for food, and that the more time the birds have to acquire skill in hunting the greater their chance of raising a chick successfully. In view of the high mortality of chicks by starvation, there could well be a threshold level of skill needed for reproductive success, below which no chick can be reared. Orians (1969) showed in the brown pelican that immature birds have only three-quarters of the success of older birds in capturing fish. Apart from this study, there is little direct evidence for Ashmole's hypothesis, although it provides an attractive explanation for postponed breeding.

The hypotheses discussed above can be examined quantitatively by means of the reproductive effort model (cf. Goodman, 1974).

We can use equations (5.56) for the adult age-classes, since the data show that adult survival and fecundity are approximately constant. We may take $\tilde{P}$ to be equal to the annual adult survival probability, P, as there is no growth of adults. Since parental care is an important component of reproductive effort, $\tilde{m}$ is taken to be one-half the average number of chicks reared per nest. The populations are probably regulated by density-dependent nestling mortality, so that $\tilde{m}$ is density-dependent and r is zero. Equations (5.56) thus give us the optimal life-history equation

$$\left|\frac{\partial P}{\partial \tilde{m}}\right| = \frac{(1-P)}{\tilde{m}} = l(b) \tag{5.61}$$

Taking $1-P$ as equal to 0.05, an increase in $\tilde{m}$ of, say, 10% of its value for the optimal life-history need only lead to a reduction in adult survival of about 0.5% in order for this relation to be satisfied. The powerful effect of high adult survival in promoting low reproductive effort is emphasised by this fact.

We must also consider the effects of increases in fecundity on survival among the juvenile age-classes, since an optimal life-history must be stable to perturbations such that reproduction occurs earlier in life. In the present case, condition (5.57) becomes

$$\left|\frac{\partial P_x}{\partial \tilde{m}_x}\right| > l(x+1) \quad (1 \leqq x < b, \tilde{m}_x = 0) \tag{5.62}$$

Since $l(x) \leqq 1$, and decreases with age, a sufficient condition for this condition to be satisfied is that the slope of survival plotted against $\tilde{m}_x$ be greater than one at the origin, for all the juvenile age-classes. The postulated increase in hunting skill with age can be incorporated into this model in the form of a decrease with age in the sensitivity of P_x to higher $\tilde{m}_x$; the sensitivity is measured by the value of $|\partial P_x/\partial \tilde{m}_x|$ for a fixed $\tilde{m}_x$. This is because a greater degree of reproductive effort will be needed to attain a given value of $\tilde{m}_x$, the lower the hunting ability of the parents. Since the survival probability of full-grown birds is high, only a modest increase in hunting skill with age is necessary to favour postponed reproduction, as may be seen by comparing equations (5.61) and (5.62).

5.4.3 *Sex differences and life-history evolution*

The work described so far has assumed that natural selection promotes similar patterns of age-specific fecundity in males and females. There are, however, some situations in which sex differences in life-history patterns are favoured by selection; two of these are reviewed briefly below. The effects of selection on the sex-ratio in age-structured populations will also be summarised.

Sex differences in timing of reproduction. In social mammals, there is frequently competition between males for mates based, in part at least, on differences in size. Large males accordingly have the greatest reproductive success (e.g. Geist, 1971, Chapter 7). If such competition is sufficiently intense, there may be strong selection on the males in favour of prolonging growth and postponing full sexual maturity (cf. section 5.3.2). Because of the competitive nature of selection in this case, the effects tend to be self-reinforcing, in the sense that selection for genotypes with prolongation of growth is intensified, once males possessing a life-history with a modest degree of postponed reproduction have been established in the population. Such selection does not act on the females, and we would therefore expect them to become fully sexually mature and cease growth considerably earlier than the males. Such sex differences have commonly been observed in social mammals where there is strong sexual competition between males (Geist, 1971; Estes, 1974).

Similar evidence for deferred maturity of males compared with females has been obtained in polygynous species of birds (Selander, 1965; Wiley, 1974a). In such species, it is frequently the case that year-old females are fully mature, while males do not acquire full breeding plumage and reproductive success until their second year. As suggested by Selander (1965), it seems likely that year-old males are relatively unsuccessful in competition with older males, due to lack of experience. It therefore pays them to acquire some experience without running the risks associated with reproduction, such as higher mortality from predators due to conspicuous breeding plumage (cf. p. 260). A correlation between polygyny and delayed maturity of males has been demonstrated in species of grackles and North American blackbirds by Selander (1965), and in grouse species by Wiley (1974a).

Sequential hermaphroditism.   This is a phenomenon whose selective causes are probably closely related to those for deferred maturity of males. It describes cases in which at least some individuals in a population start reproductive life as one sex, and change over to the other sex at some point. Sequential hermaphrodites may be *protandrous* or *protogynous*, depending on whether they start life as males or females, respectively. It is common, for example, among certain groups of fishes, such as the wrasses. Ghiselin (1969) suggested that sequential hermaphroditism has a selective advantage when there is a tendency for the fecundity of one sex to increase more rapidly with age than that of the other. Consider, for example, the case of an initially bisexual species with two age-classes, and in which the age-specific survival probabilities for males and females are equal. If the population size is stationary, then the fact that each individual has a mother and a father implies that, using the notation of section 1.2.1, we must have

$$l(1)M(1) + l(2)M(2) = l(1)M^*(1) + l(2)M^*(2) \qquad (5.63)$$

If female fecundity is constant with age but male fecundity increases, we have $M(1) = M(2)$ and $M^*(1) < M^*(2)$. It follows from equation (5.63) that $M^*(1) < M(1) < M^*(2)$. The carriers of a rare mutant gene which causes them to be female up to age-class 1, and male in age-class 2, will have an expectation of offspring of $l(1)M(1) + l(2)M^*(2)$, and so will be at a selective advantage over the rest of the population. This problem has been investigated in more general mathematical terms by Warner (1975), Warner, Robertson and Leigh (1975) and Leigh, Charnov and Warner (1976).

On this theory, protogyny is especially likely to evolve when there is sexual competition between males based on size. Warner *et al.* (1975) give a detailed account of protogyny in the blue wrasse, *Thalassoma bifasciatum*. They show that the fecundity of the females, as measured by ovary weight, increases only slightly more rapidly than body weight with increasing size, whereas only the largest males appear to be able to hold breeding territories. Sex-reversed males are nearly always of the larger and more brightly coloured class of males, who are most successful in breeding.

Selection and the sex-ratio. Fisher (1930, pp. 158–60) was the first to propose a mechanism for the adjustment of the sex-ratio by natural selection. He pointed out that both parents make an equal genetic contribution to a zygote, with respect to autosomal loci. A male zygote will therefore have a higher net fitness (survival × fecundity) than a female zygote, in a discrete-generation population whose primary sex-ratio (sex-ratio among the zygotes, section 1.2.1) is biased towards females. (This is true even if males have a lower probability of survival than females.) Consequently, a geno-type which tends to produce offspring with a higher frequency of male zygotes than average, but otherwise has the same fecundity and survival probability, will have a selective advantage. If the sex-ratio is biased in favour of males, then female-producing geno-types are favoured. The sex-ratio should therefore tend to stabilise at one-half, if there is genetic variation available at autosomal loci. This idea has been confirmed by studies of one-locus models of sex-ratio variation, e.g. Spieth (1974). Variation at X-linked loci has similar consequences (Trivers and Hare, 1976).

As Fisher realised, a primary sex-ratio of one-half is not neces-sarily favoured if parents can replace offspring which die before reaching a certain age. If, for example, males tend to have a higher mortality than females, then a mother who produces a higher frequency of male zygotes than average will compensate for the greater frequency of death of these males by producing more zygotes than average, so that she will have an increased fecundity over and above any advantage due to a female-biased population sex-ratio. This results in a primary sex-ratio biased in favour of males at equilibrium. It can be shown that selection favours a sex-ratio of one-half at the age at which parents can no longer compensate for death of an offspring (Crow and Kimura, 1970, pp. 291–2).

All these results apply to discrete-generation populations. The problem of modelling sex-ratio evolution in an age-structured population is a complex one, but a beginning has been made by Leigh (1970), Leigh *et al.* (1976) and Charlesworth (1977), who examined the conditions for spread of a rare gene affecting the primary sex-ratio of the offspring of its carriers. This enables a determination of the ESS sex-ratio, i.e. that sex-ratio which does not permit the spread of any mutant gene affecting sex-ratio (cf. p. 231). Charlesworth's model allowed for possible dependence of the sex-ratio of offspring upon the age as well as the genotype of the parents. He showed that the ESS is a sex-ratio of one-half among the zygotes produced by the population as a whole, when there is no compensation by parents for the loss of offspring or no sex-difference in mortality. No particular pattern of primary sex-ratio with respect to parental age is favoured under these conditions, provided that the overall primary sex-ratio is one-half.

Charlesworth also studied models in which there was compensation of the type likely to occur in non-contracepting human populations. Here, there is a prolonged period of natural sterility during lactation, so that a child which dies *in utero* or before weaning causes its mother to become fertile sooner than if the child had survived to weaning. A maternal genotype that tends to produce a higher-than-average frequency of male zygotes (which have a higher mortality than females during most of the period of life in question) is thus associated with a shorter interval between successive conceptions than average. Using the renewal model of Henry (1953, 1957) for human age-specific fecundity, Charlesworth showed that the ESS primary sex-ratio for human populations is biased in favour of males, as expected from the discrete-generation result. By the end of the first eighteen months or so of post-natal life, death of an infant is not likely to shorten the period of infertility significantly, and the ESS sex-ratio at this age is biased in favour of females. If the mortality rates for male and female infants in the first year of life in typical pre-industrial populations are combined with typical statistics on sex-ratios at birth (which are usually male-biased), a female-biased sex-ratio for one-year old infants is obtained, in agreement with the ESS prediction.

It was also possible to carry out a somewhat crude analysis of the ESS for sex-ratio as a function of parental age, as well as the overall sex-ratio. A tendency for younger females to produce a higher frequency of males than older females was predicted. An intuitive interpretation of this is that older women have a greater chance of becoming sterile or dying in the intervals between conceptions than young women, so that they have a lower net ability to compensate for the deaths of offspring than younger women. Their offspring should accordingly have a sex-ratio closer to one-half. It is interesting to note that the sex-ratio at birth in human populations is dependent upon maternal age in the way predicted by this theory (Novitski and Kimball, 1958; Teitelbaum, 1972). There is thus reasonably good agreement between the expectations of sex-ratio theory and observations on human populations. This should not be taken as confirming the theory, since sex-ratios at birth are affected by a variety of socio-economic factors, and alternative explanations of the association between maternal age and sex-ratio can be put forward. If it could be shown that the primary sex-ratio in man is indeed biased in favour of males, the theory would be considerably strengthened, but at present this sex-ratio is an unknown quantity (Charlesworth, 1977).

5.4.4 *Some general aspects of life-history theory*
We shall conclude this chapter with a brief discussion of some aspects of life-history theory that do not fit easily into the above categories.

The uses of reproductive value. Ever since Fisher introduced the concept in 1930, there has been a strong temptation for biologists to use reproductive value as a tool for analysing life-history evolution and related problems. As we have seen, it cannot be used as an index of the intensity of selection as a function of age, despite Fisher's (1930, p. 29) remarks concerning the possible significance of the inverse relationship between reproductive value and mortality observable in human populations, and Medawar's use of it in discussing the evolution of senescence (section 5.2.2). Reproductive value or, rather, Schaffer's (1974*a*) generalisation of it, does play an important role in determining the conditions for an optimal life-history under the reproductive effort model (section 5.3.2). Furthermore, as discussed by Schaffer and by Taylor *et al.* (1974), there is a certain sense in which reproductive value at each age is maximised in optimal life-histories. This may be seen as follows. As we saw on p. 232, the condition for an optimal life-history in a density-independent population is that the sum $\sum e^{-rx}k(x)$ be at a local maximum with respect to perturbations in $k(x)$, r being held constant at the value which makes the original sum unity. If we consider a certain age y, variations in the life-history at age y and later ages affect only the component $\sum_{x=y} e^{-rx}k(x)$ of the above sum, regardless of the particular constraints assumed in the optimisation model. Hence, if a given life-history maximises $\sum e^{-rx}k(x)$, the sum $\sum_{x=y} e^{-rx}k(x)$ must be maximised with respect to perturbations at ages y, $y+1$, ..., the life-history at earlier ages being held constant. This is true for each value of y. Conversely, maximisation of each sum $\sum_{x=y} e^{-rx}k(x)$ implies that the life-history is optimal. The maximisation of reproductive value follows from this, since we have

$$v(y) = \frac{e^{ry}}{l(y)} \sum_{x=y} e^{-rx}k(x)$$

and $l(y)$ is unaffected by perturbations at ages y and beyond. Similar remarks apply to density-dependent populations. There is, of course, nothing of particular biological significance in this maximisation of reproductive value, since *any* product of $\sum_{x=y} e^{-rx}k(x)$ and a factor that is independent of events at age y and beyond would have the same properties.

The restriction that r be held constant while carrying out perturbations to the life-history can be relaxed for the case of perturbations to age-classes which do not have boundary values for the life-history variables under consideration (e.g. age-classes with intermediate reproductive effort). The total differential of r, dr, must be zero at a local maximum in r with respect to such perturbations, so that the restriction of constant r when finding maxima in reproductive value etc. is unnecessary in this case. With the reproductive effort model, described in section 5.3.2, $v(y)$ (or, more generally, $\tilde{v}(y)$, if growth as well as survival is taken into account) is maximised with respect to perturbations at ages $x < y$ as well as $x \geq y$. (The restriction of constant r must be applied if these ages have reproductive efforts of 0 or 1.) This is because $P(x)$ or $\tilde{P}(x)$ depend only on $m(x)$ or $\tilde{m}(x)$ in this model, so that perturbations to the life-history at ages $x < y$ leave $v(y)$ or $\tilde{v}(y)$ unaffected, since $P(x)$ or $\tilde{P}(x)$ for $x < y$ can be cancelled in the expressions for reproductive value at age y. The result is, of course, not true if the effects of changes in reproductive effort at ages $x < y$ influence survival or growth at age y and beyond. There is therefore a much stronger sense in which selection can be said to maximise reproductive value in reproductive effort models, than in more general optimisation models.

Another role for reproductive value has been suggested by Slobodkin (1974). He proposed that the life-history of a prey species which is subjected to an increased rate of predation at a certain age would tend to evolve in such a way that reproductive value is withdrawn maximally from that age. This possibility has been investigated analytically by Michod (1979), using the methods described on p. 248. He was able to confirm Slobodkin's suggestion for certain types of population, although its general validity remains to be established.

Finally, MacArthur and Wilson (1967, Chapter 4) suggested that a colonising species would be selected to send out individuals to found new colonies at an age when these individuals have maximal reproductive value, on the grounds that their probability of successfully founding a new population would be highest at this age. This was investigated by Williamson and Charlesworth (1976), using branching process theory. They found that there is often no particular relationship between reproductive value at a given age and the probability of survival of a population founded by an individual of that age.

Maximisation of population parameters.   A number of life-history studies have been published which make use of the maximisation of population parameters. For example, Holgate (1967) and Mountford (1971, 1973) treated certain problems by finding the maximum probability of survival of a population. Cohen (1966, 1967, 1968, 1970) maximised the geometric mean of the growth-rate of a population in a temporally varying environment. As has already been discussed (section 5.1), it is necessary to demonstrate that natural selection can actually have the effect of maximising the population parameter in question, unless one is willing to invoke group selection. The use of survival probability can be justified in selectionist terms by regarding the calculations in question as dealing with the probability of survival of a mutant gene rather than a whole population, as mentioned in section 5.2.1. Although the calculations often turn out to be identical, the conceptual distinction between the two procedures is important. Similarly, Cohen's models can be translated into considerations of the rate of spread of a rare gene, along the lines indicated in section 4.3.3.

Are life-histories optimal? Two different methods for the theoretical study of life-history evolution have been presented in this chapter. In the first section, the phenomenon of senescence was treated as being due, in a sense, to the failure of perfect adaptation, since selection is more effective in changing fitness parameters early in life. In the second part, the life-history was regarded as being close to an optimum, at which fitness is maximised according to some set of constraints whose form is not itself subject to selective change. The relative importance of these two aspects of life-history evolution is a matter for empirical investigation. Since organisms live in a changing environment, and it is unlikely that exactly the right kind of genetic variability will always be available at the right time, any theory which treats present-day life-histories as optimal must be approximate. Since selection is more effective early in life, we might well expect to find a closer correspondence between real and optimal life-histories early rather than late in life. Another difficulty with optimal theory is knowing what constraints are likely to be relevant. Although qualitative agreement between theoretical expectation and observed differences in life-histories between and within species is often good, no studies of natural populations have really been capable of demonstrating close quantitative agreement between a real and an optimal life-history. Further careful studies of natural populations may improve on this situation.

APPENDIX 1

Generating functions and their properties

The purpose of this appendix is to summarise the properties of generating functions used in this book. For a more detailed treatment, Feller (1968, Chapter 11) should be consulted.

Consider a finite or infinite sequence of real numbers $a(0), a(1), a(2), \ldots$. The *generating function* of this sequence, $\bar{a}(s)$, is defined by the function

$$\bar{a}(s) = \sum_{j=0}^{\infty} s^j a(j) \tag{A1.1}$$

where s is a real or complex variable; $\bar{a}(s)$ is defined only for values of s for which the sum converges. For a finite sequence of $n + 1$ numbers, the upper limit of the summation is taken as n. If the sequence a defines a probability distribution, then $\bar{a}(s)$ is called a *probability generating function*, for obvious reasons. It is easily seen that the mean of the distribution is given by $(d\bar{a}(s)/ds)_{s=1}$; more generally, $(d^r\bar{a}(s)/ds^r)_{s=1}$ gives the rth factorial moment of the distribution.

One very useful property of generating functions is exploited in the manipulation of the *convolution*, c, of two sequences a and b. The convolution is itself a sequence, whose jth element $c(j)$ is defined by

$$c(j) = \sum_{k=0}^{j} a(k)b(j-k) \tag{A1.2a}$$

It is easily seen that the generating function of c is given by the following simple expression

$$\bar{c} = \bar{a}\bar{b} \tag{A1.2b}$$

Another useful property of generating functions arises from the fact that they can often be expanded into partial fractions. Suppose

269

we have a generating function $\bar{a}$ of the form

$$\bar{a}(s) = \bar{b}(s)/\bar{d}(s) \tag{A1.3}$$

where $\bar{b}$ and $\bar{d}$ are both polynomials in s without common zeros. Assume for simplicity that $\bar{b}$ is of lower degree than $\bar{d}$, and that $\bar{d}$ has m distinct zeros, so that we can write

$$\bar{d}(s) = (s - s_1)(s - s_2) \ldots (s - s_m) \tag{A1.4}$$

It is known from algebra that in this case $\bar{a}$ can be expanded into partial fractions

$$\bar{a}(s) = \sum_{j=1}^{m} \frac{C_j}{(s_j - s)} \tag{A1.5}$$

where the C_j are constants. The value of C_1 can be found by multiplying both sides of equation (A1.5) by $s_1 - s$, and letting s approach s_1. The product $(s_1 - s)\bar{a}(s)$ approaches C_1 as $s \to s_1$; from equations (A1.3) and (A1.4), we have

$$(s_1 - s)\bar{a}(s) = \frac{-\bar{b}(s)}{(s - s_2)(s - s_3) \ldots (s - s_m)}$$

As $s \to s_1$, the numerator approaches $\bar{b}(s_1)$ and the denominator tends to $(s_1 - s_2)(s_1 - s_3) \ldots (s_1 - s_m)$, which is equal to $(d\bar{d}/ds)_{s=s_1}$. A similar procedure applied to each zero in turn yields the general expression

$$C_j = \frac{-\bar{b}(s_j)}{(d\bar{d}/ds)_{s=s_j}} \tag{A1.6}$$

We also have

$$\frac{1}{(s_j - s)} = \frac{1}{s_j(1 - s/s_j)} = \frac{1}{s_j}\left[1 + \frac{s}{s_j} + \left(\frac{s}{s_j}\right)^2 + \left(\frac{s}{s_j}\right)^3 + \ldots\right]$$

provided that $|s| < |s_j|$. Substituting from this into equation (A1.5) and equating coefficients of s on both sides, we obtain the coefficient of s^k

$$a(k) = \sum_{j=1}^{m} \frac{C_j}{s_j^{k+1}} \tag{A1.7}$$

If s_1 is a zero of $\bar{d}$ which is smaller in modulus than any other, then as $k \to \infty$, we obtain the asymptotic expression

$$a(k) \sim \frac{C_1}{s_1^{k+1}} \qquad (A1.8)$$

Relatively trivial modifications are required to deal with cases when $\bar{b}$ is of the same or greater degree than $\bar{d}$, or when $\bar{d}$ has repeated zeros. The asymptotic expression (A1.8) still holds in these cases, provided that s_1 is a unique zero; a related expression can be obtained even when s_1 is a repeated zero (Feller, 1968, p. 285).

APPENDIX 2

Asymptotic values of Δp_i and $\Delta^2 p_i$

Equation (4.1) gives the number of births at time t. It can be written as

$$B(t) = \sum_x B(t-x)[k_s(x) + O(\varepsilon)] \qquad (A2.1)$$

where the term $O(\varepsilon)$ depends on the gene frequencies at time $t - x$ and on the genotypic differences in the reproductive functions. It follows that it must be possible to choose two positive numbers of order ε, ε_1 and ε_2, such that $k_s(x) - \varepsilon_1 < k_s(x) + O(\varepsilon) < k_s(x) + \varepsilon_2$, for all values of t. It can be shown that

$$\sum_x B_1(t-x)[k_s(x) - \varepsilon_1] < B(t) < \sum_x B_2(t-x)[k_s(x) + \varepsilon_2] \qquad (A2.2)$$

where $B_1(t)$ and $B_2(t)$ are the solutions to the renewal equations associated with the reproductive functions $k_s(x) - \varepsilon_1$ and $k_s(x) + \varepsilon_2$ (the initial conditions for B_1 and B_2 are assumed to be the same as for B). Using the standard solution to the renewal equation, given by (1.39), for sufficiently large t we have

$$B_1(t) \sim C_1 e^{(r_s - \eta_1)t} < B(t) < B_2(t) \sim C_2 e^{(r_s + \eta_2)t} \qquad (A2.3)$$

where $r_s - \eta_1$ and $r_s + \eta_2$ are the intrinsic rates of increases associated with the reproductive functions $k_s(x) - \varepsilon_1$ and $k_s(x) + \varepsilon_2$. Both η_1 and η_2 are $O(\varepsilon)$, as is $C_1 - C_2$. Equation (A2.3) implies that

$$\Delta \ln B \sim r_s + O(\varepsilon) \qquad (A2.4)$$

Exactly the same argument can be applied to equation (4.5). We have

$$B(t)p_i(t) = \sum_x B(t-x)p_i(t-x)[k_s(x) + O(\varepsilon)] \qquad (A2.5)$$

272

As in thé previous case, positive numbers of order ε, ε_3 and ε_4, can be assigned, such that (for sufficiently large t) we have

$$C_3\, e^{(r_s-\eta_3)t} < B(t)p_i(t) < C_4\, e^{(r_s-\eta_4)t} \tag{A2.6}$$

where η_3 and η_4 are defined analogously to η_1 and η_2, and are equal to $r_s + O(\varepsilon)$. We therefore obtain the asymptotic expression

$$\Delta \ln Bp_i \sim r_s + O(\varepsilon) \tag{A2.7}$$

Subtracting equation (A2.4) from equation (A2.7), we find that

$$\Delta \ln p_i \sim O(\varepsilon) \tag{A2.8a}$$

from which we obtain

$$\Delta p_i \sim O(\varepsilon) \tag{A2.8b}$$

Having established that Δp_i is asymptotically of order ε, we can now consider the magnitude of $\Delta^2 p_i$, the second difference in p_i. The $O(\varepsilon)$ term in equation (A2.8b) for a given time t is a complicated function, ϕ_i say, of the sequences of $O(\varepsilon)$ terms in equations (A2.1) and (A2.5) at successive times t, $t-1, \ldots$. Similarly, Δp_i for the preceding time $t-1$ is a function, ϕ_i^*, of the $O(\varepsilon)$ terms for $t-1, t-2, \ldots$. At time $t-1$ we have $\Delta^2 p_i = \phi_i - \phi_i^*$, which is of the same order as the first differences in the $O(\varepsilon)$ terms in equations (A2.1) and (A2.5). These terms involve only the products of gene frequencies and *constant* $O(\varepsilon)$ terms given by the genotypic differences in reproductive functions. Without loss of generality, we can assume that t is taken sufficiently large that the gene frequency terms are changing by $O(\varepsilon)$, so that the first differences in the $O(\varepsilon)$ terms are $O(\varepsilon^2)$. This establishes that $\Delta^2 p_i$ is asymptotically $O(\varepsilon^2)$.

This result can be used to establish the approximation for $p_i(t)$ in terms of $p_i(t-x)$ and $\Delta p_i(t)$, used on p. 157. We have

$$p_i(t) = p_i(t-x) + \sum_{y=1}^{x} \Delta p_i(t-y) \tag{A2.9}$$

Furthermore, for $y \geqq 1$ we have (writing $\Delta^2 p_i(t) = \Delta p_i(t+1) - \Delta p_i(t)$)

$$\Delta p(t) = \Delta p(t-y) + \sum_{u=1}^{y} \Delta^2 p(t-u) \tag{A2.10}$$

But $\Delta^2 p_i$ is asymptotically $O(\varepsilon^2)$, so that, combining equations (A2.9) and (A2.10), we obtain

$$p_i(t) \sim p_i(t-x) + x\Delta p_i(t) + O(\varepsilon^2) \qquad (A2.11)$$

These asymptotic results are based on the properties of the renewal equation. The period of time needed for convergence to the asymptotic solution of this equation is of the order of a few 'generations' in biological applications, as discussed in section 1.3.2. The above results should, therefore, be valid after a relatively short time has elapsed.

REFERENCES

Anderson, W. W. 1971. Genetic equilibrium and population growth under density-regulated selection. *Am. Nat.*, **105**, 489–98.

Anderson, W. W. & C. E. King. 1970. Age-specific selection. *Proc. Natl. Acad. Sci. USA*, **66**, 780–6.

Anderson, W. W. & T. K. Watanabe. 1980. A demographic approach to selection on *Drosophila pseudoobscura* karyotypes. In preparation.

Ashmole, N. P. 1963. The regulation of numbers of oceanic birds. *Ibis*, **103b**, 458–73.

Aspinwall, N. 1974. Genetic analysis of North American populations of the pink salmon *Oncorhynchus gorbuscha*, possible evidence for the neutral mutation–random drift hypothesis. *Evolution*, **28**, 295–305.

Auslander, D., G. F. Oster & C. Huffaker. 1974. Dynamics of interacting populations. *J. Franklin Inst.*, **297**, 345–75.

Bakker, K. 1959. Feeding period, growth and pupation in larvae of *Drosophila melanogaster*. *Entomol. Exp. Appl.*, **2**, 171–86.

Barkalow, F. S., R. B. Hamilton & R. F. Soots, 1970. The vital statistics of an unexploited gray squirrel population. *J. Wildl. Manage.*, **34**, 489–500.

Beardmore, J. A., T. Dobzhansky & O. Pavlovsky. 1960. An attempt to compare the fitness of polymorphic and monomorphic experimental populations of *Drosophila pseudoobscura*. *Heredity*, **14**, 19–33.

Beddington, J. R. 1974. Age-distribution and the stability of simple discrete time population models. *J. Theor. Biol.*, **47**, 65–74.

Beddington, J. R. & C. A. Free. 1976. Age structure effects in predator-prey interactions. *Theor. Popul. Biol.*, **9**, 15–24.

Bell, G. 1976. On breeding more than once. *Am. Nat.*, **110**, 57–77.

Bernadelli, H. 1941. Population waves. *J. Burma Res. Soc.*, **31**, 1–18.

Bodmer, W. F. 1965. Differential fertility in population genetics. *Genetics*, **51**, 411–42.

Bodmer, W. F. 1968. Demographic approaches to the measurement of differential selection in human populations. *Proc. Natl. Acad. Sci. USA*, **59**, 690–9.

Botkin, D. B. & R. S. Miller. 1974. Mortality rates and survival of birds. *Am. Nat.*, **108**, 181–92.

Bryant, E. 1971. Life-history consequences of natural selection: Cole's result. *Am. Nat.*, **105**, 75–6.

Bulmer, M. G. 1977. Periodical insects. *Am. Nat.*, **111**, 1099–1117.

Bulmer, M. G. & C. M. Perrins. 1973. Mortality in the great tit, *Parus major*. *Ibis*, **115**, 277–81.

Buzzati-Traverso, A. A. 1955. Evolutionary changes in components of fitness and other polygenic traits in *Drosophila melanogaster* populations. *Heredity*, **9**, 153–86.

Caughley, G. 1966. Mortality in mammals. *Ecology*, **47**, 906–17.

Caughley, G. 1977. *Analysis of Vertebrate Populations*. New York, London: Wiley.

Cavalli–Sforza, L. L. & W. F. Bodmer. 1971. *The Genetics of Human Populations*. San Francisco: W. H. Freeman.

Cavalli-Sforza, L. L., M. Kimura & I. Barrai. 1966. The probability of consanguineous marriages. *Genetics*, **54**, 37–60.

Charlesworth, B. 1970. Selection in populations with overlapping generations. I. The use of Malthusian parameters in population genetics. *Theor. Popul. Biol.*, **1**, 352–70.

Charlesworth, B. 1971. Selection in density-regulated populations. *Ecology*, **52**, 469–74.

Charlesworth, B. 1972. Selection in populations with overlapping generations. III. Conditions for genetic equilibrium. *Theor. Popul. Biol.*, **3**, 377–95.

Charlesworth, B. 1973. Selection in populations with overlapping generations. V. Natural selection and life histories. *Am. Nat.*, **107**, 303–11.

Charlesworth, B. 1974a. Selection in populations with overlapping generations. VI. Rates of change of gene frequency and population growth rate. *Theor. Popul. Biol.*, **6**, 108–32.

Charlesworth, B. 1974b. The Hardy–Weinberg law with overlapping generations. *Adv. Appl. Probab.*, **6**, 4–6.

Charlesworth, B. 1976. Natural selection in age-structured populations. In *Lectures on Mathematics in the Life Sciences*, vol. 8, ed. S. Levin, pp. 69–87. Providence, R.I.: American Mathematical Society.

Charlesworth, B. 1977. Population genetics, demography and the sex ratio. In *Measuring Selection in Natural Populations*, ed. F. B. Christiansen and T. M. Fenchel, pp. 345–63. Berlin: Springer–Verlag.

Charlesworth, B. & D. Charlesworth. 1973. The measurement of fitness and mutation rate in human populations. *Ann. Hum. Genet.*, **37**, 175–87.

Charlesworth, D. & B. Charlesworth. 1975. Theoretical genetics of Batesian mimicry. I. Single-locus models. *J. Theor. Biol.*, **55**, 283–303.

Charlesworth, B. & J. T. Giesel. 1972a. Selection in populations with overlapping generations. II. Relations between gene frequency and demographic variables. *Am. Nat.*, **106**, 388–401.

Charlesworth, B. & J. T. Giesel. 1972b. Selection in populations with overlapping generations. IV. Fluctuations in gene frequency with density-dependent selection. *Am. Nat.*, **106**, 402–11.

Charlesworth, B. & J. A. León. 1976. The relation of reproductive effort to age. *Am. Nat.*, **110**, 449–59.

Charlesworth, B. & J. A. Williamson. 1975. The probability of survival of a mutant gene in an age-structured population and implications for the evolution of life-histories. *Genet. Res.*, **26**, 1–10.

Charnov, E. L. & W. M. Schaffer. 1973. Life-history consequences of natural selection: Cole's result revisited. *Am. Nat.*, **107**, 791–3.

Chitty, D. 1960. Population processes in the vole and their relevance to general theory. *Can. J. Zool.*, **38**, 99–113.

Choy, S. C. & B. S. Weir. 1978. Exact inbreeding coefficients in populations with overlapping generations. *Genetics*, **89**, 591–614.

Clarke, J. M., J. Maynard Smith & K. C. Sondhi. 1961. Asymmetrical response to selection for rate of development in *Drosophila subobscura*. *Genet. Res.*, **2**, 70–81.

Coale, A. J. 1957. How the age distribution of a human population is determined. *Cold Spring Harbor Symp. Quant. Biol.*, **22**, 83–9.

Coale, A. J. 1972. *The Growth and Structure of Human Populations*. Princeton: Princeton University Press.

Cohen, D. 1966. Optimising reproduction in a randomly varying environment. *J. Theor. Biol.*, **12**, 110–29.

Cohen, D. 1967. Optimising reproduction in a randomly varying environment when a correlation may exist between the conditions at the time a choice has to be made and the subsequent outcome. *J. Theor. Biol.*, **16**, 1–14.

Cohen, D. 1968. A general model of optimal reproduction. *J. Ecol.*, **56**, 219–28.

Cohen, D. 1970. A theoretical model for the optimal timing of diapause. *Am. Nat.*, **104**, 389–400.

Cohen, J. E. 1976. Ergodicity of age structure in populations with Markovian vital rates. I. Countable states. *J. Am. Stat. Assoc.*, **71**, 335–9.

Cohen, J. E. 1977a. Ergodicity of age structure in populations with Markovian vital rates. II. General states. *Adv. Appl. Probab.*, **9**, 18–37.

Cohen, J. E. 1977b. Ergodicity of age structure in populations with Markovian vital rates. III. Finite-state moments and growth rate: an illustration. *Adv. Appl. Probab.*, **9**, 462–75.

Cole, L. C. 1954. The population consequences of life history phenomena. *Q. Rev. Biol.*, **29**, 103–37.

Comfort, A. 1979. *The Biology of Senescence*, 3rd edn. Edinburgh, London: Churchill Livingstone.

Cornette, J. L. 1975. Some basic elements of continuous selection models. *Theor. Popul. Biol.*, **8**, 301–13.

Cornette, J. L. 1978. Sex-linked genes in an age-structured population. *Heredity*, **40**, 291–7.

Crow, J. F. 1978. Gene frequency and fitness change in an age-structured population. *Ann. Hum. Genet.*, **42**, 335–70.

Crow, J. F. & M. Kimura. 1970. *An Introduction to Population Genetics Theory*. New York: Harper & Row.

Crow, J. F. & M. Kimura. 1971. The effective number of a population with overlapping generations: a correction and further discussion. *Am. J. Hum. Genet.*, **24**, 1–10.

Dahlberg, G. 1929. Inbreeding in man. *Genetics*, **14**, 421'–54.

Dahlberg, G. 1948. *Mathematical Methods for Population Genetics*. Basle: S. Karger.

Deevey, E. S. 1947. Life tables for natural populations. *Q. Rev. Biol.*, **22**, 283–314.

Demetrius, L. 1969. The sensitivity of population growth rate to perturbations in the life-cycle components. *Math. Biosci.*, **4**, 129–36.

Dickerson, G. E. & L. N. Hazel. 1944. Effectiveness of selection on progeny performance as a supplement to earlier culling of livestock. *J. Agric. Res.*, **59**, 459–76.

Dobzhansky, T. 1943. Genetics of natural populations. IX. Temporal changes in the composition of populations of *Drosophila pseudoobscura*. *Genetics*, **28**, 162–86.

Dobzhansky, T. & H. Levene. 1951. Development of heterosis through natural selection in populations of *Drosophila pseudoobscura*. *Am. Nat.*, **85**, 247–264.

Dubinin, N. P. & G. G. Tiniakov. 1945. Seasonal cycles and the concentration of inversions in populations of *Drosophila funebris*. *Am. Nat.*, **79**, 570–2.

Du Mouchel, W. H. & W. W. Anderson. 1968. The analysis of selection in experimental populations. *Genetics*, **58**, 435–49.

Elsen, J. M. & J. C. Macquot. 1974. *Recherches pour une Rationalisation Technique et Économique des Schémas de Sélection des Bovins et Ovins. Bull. Tech. Dep. Génét. Anim.* No. 17. Paris: Jouy-en-Josas, Institut National de la Recherche Agronomique.

Emigh, T. H. 1979a. The dynamics of finite haploid populations with overlapping generations. I. Moments, fixation probabilities and stationary distributions. *Genetics*, **92**, 323–37.

Emigh, T. H. 1979b. The dynamics of finite haploid populations with overlapping generations. II. The diffusion approximation. *Genetics*, **92**, 338–51.

Emigh, T. H. & E. Pollak. 1979. Fixation probabilities and effective population numbers in diploid populations with overlapping generations. *Theor. Popul. Biol.*, **15**, 86–107.

Emlen, J. M. 1970. Age-specificity and ecological theory. *Ecology*, **51**, 588–601.

Estes, R. D. 1974. Social organisation of the African Bovidae. In *The Behaviour of Ungulates and its Relation to Management*, ed. V. Geist and F. Walther, pp. 166–205. Morges, Switzerland: IUCN.

Euler, L. 1760. Recherches generales sur la mortalité et la multiplication. *Mémoires de l'Académie Royale des Sciences et Belles Lettres*, **16**, 144–64. (Translated by N. Keyfitz & B. Keyfitz (1970), *Theor. Popul. Biol.*, **1**, 307–14.)

Ewens, W. J. 1969. *Population Genetics*. London: Methuen.

Fagen, R. M. 1972. An optimal life history in which reproductive effort decreases with age. *Am. Nat.*, **104**, 258–61.

Falconer, D. S. 1960. *An Introduction to Quantitative Genetics*. Edinburgh: Oliver & Boyd.

Feller, W. 1941. On the integral equation of renewal theory. *Ann. Math. Stat.*, **12**, 243–67.

Feller, W. 1966. *An Introduction to Probability Theory and its Applications*, vol. 2. New York, London: Wiley.

Feller, W. 1968. *An Introduction to Probability Theory and its Applications*. vol. 1, 3rd edn. New York, London: Wiley.

Felsenstein, J. 1971. Inbreeding and variance effective numbers in populations with overlapping generations. *Genetics*, **68**, 581–97.

Festing, M. F. W. & D. K. Blackmore. 1971. Lifespan of pathogen-free (MRC category 4) mice and rats. *Lab. Anim.*, **5**, 179–92.

Fisher, R. A. 1922. On the dominance ratio. *Proc. R. Soc. Edinburgh*, **42**, 321–41.

Fisher, R. A. 1928a. The possible modification of the response of the wild type to recurrent mutations. *Am. Nat.*, **62**, 115–26.

Fisher, R. A. 1928*b*. Two further notes on the evolution of dominance. *Am. Nat.*, **62**, 571–4.

Fisher, R. A. 1930. *The Genetical Theory of Natural Selection.* Oxford: Clarendon Press. (Reprinted and revised, 1958. New York: Dover.)

Fisher, R. A. 1941. Average excess and average effect of a gene substitution. *Ann. Eugen.*, **11**, 53–63.

Frota-Pessoa, O. 1957. The estimation of the size of isolates based on census data. *Am. J. Hum. Genet.*, **2**, 9–16.

Gadgil, M. & W. H. Bossert. 1970. Life history consequences of natural selection. *Am. Nat.*, **102**, 52–64.

Gaines, M. S. & C. J. Krebs. 1971. Genetic changes in fluctuating vole populations. *Evolution*, **25**, 702–23.

Gaines, M. S., L. R. McCleaghan & R. K. Rose. 1978. Temporal patterns of allozymic variation in fluctuating populations of *Microtus ochrogaster*. *Evolution*, **32**, 723–9.

Geist, V. 1971. *Mountain Sheep: a Study in Evolution and Behaviour.* Chicago: University of Chicago Press.

Gerking, S. D. 1959. Physiological changes accompanying ageing in fishes. In *The Life-span of Animals. CIBA Foundation Colloquia on Ageing*, vol. 5, ed. G. E. Wolstenholme and M. O'Connor, pp. 181–207. London: Churchill.

Gershenson, S. 1945. Evolutionary studies on the distribution and dynamics of melanism in the hamster (*Cricetus cricetus*). II. Seasonal and annual changes in the frequency of black hamsters. *Genetics*, **30**, 233–51.

Ghiselin, M. T. 1969. The evolution of hermaphroditism among animals. *Q. Rev. Biol.*, **44**, 189–208.

Giesel, J. T. 1971. The relations between population structure and rate of inbreeding. *Evolution*, **25**, 491–6.

Gillespie, J. H. 1973*a*. Polymorphism in random environments. *Theor. Popul. Biol.*, **4**, 193–5.

Gillespie, J. H. 1973*b*. Natural selection with varying selection coefficients–a haploid model. *Genet. Res.*, **21**, 115–20.

Gillespie, J. H. 1978. A general model to account for enzyme variation in natural populations. V. The SAS/CFF model. *Theor. Popul. Biol.*, **14**, 1–45.

Goel, N. S. & M. Ycas. 1976. The error catastrophe theory and ageing. *J. Math. Biol.*, **3**, 121–47.

Gonzalez, B. M. 1923. Experimental studies on the duration of life. VIII. The influence on duration of life of certain mutant genes of *Drosophila melanogaster*. *Am. Nat.*, **57**, 289–325.

Goodman, L. A. 1967. On the reconciliation of mathematical theories of population growth. *J. R. Stat. Soc.*, **A130**, 541–53.

Goodman, L. A. 1969. The analysis of population growth when birth and death rates depend upon several factors. *Biometrics*, **25**, 660–5.

Goodman, L. A. 1971. On the sensitivity of the intrinsic growth rate to changes in the age-specific birth and death rates. *Theor. Popul. Biol.*, **2**, 339–54.

Goodman, D. 1974. Natural selection and a cost ceiling on reproductive effort. *Am. Nat.*, **108**, 247–68.

Gowen, J. W. & L. E. Johnson. 1946. On the mechanism of heterosis. I. Metabolic capacity of different races of *Drosophila melanogaster* for egg production. *Am. Nat.*, **80**, 149–79.

Gregorius, H-R. 1976. Convergence of genetic compositions assuming infinite population size and overlapping generations. *J. Math. Biol.*, **3**, 179–186.

Guckenheimer, J., G. F. Oster & A. Ipaktchi. 1977. The dynamics of density-dependent population models. *J. Math. Biol.*, **4**, 101–47.

Gurtin, M. E. & R. C. MacCamy. 1974. Non-linear age-dependent dynamics. *Arch. Ration. Mech. Anal.*, **54**, 281–300.

Hairston, N. G., D. W. Tinkle & H. M. Wilbur. 1970. Natural selection and the parameters of population growth. *J. Wildl. Manage.*, **34**, 681–90.

Hajnal, J. 1963. Concepts of random mating and the frequency of consanguineous marriages. *Proc. R. Soc. Lond.*, **B159**, 125–77.

Hajnal, J. 1976. The absolute numbers of consanguineous marriages. *Adv. Appl. Probab.*, **8**, 630–5.

Haldane, J. B. S. 1927*a*. A mathematical theory of natural and artificial selection. Part IV. *Proc. Camb. Phil. Soc.*, **23**, 607–15.

Haldane, J. B. S. 1927*b*. A mathematical theory of natural and artificial selection. Part V. Selection and mutation. *Proc. Camb. Phil. Soc.*, **23**, 838–44.

Haldane, J. B. S. 1941. *New Paths in Genetics*. London: Allen and Unwin.

Haldane, J. B. S. 1953. Animal populations and their regulation. In *Penguin New Biology*, vol. 15, pp. 9–24. London: Penguin.

Haldane, J. B. S. 1962. Natural selection in a population with annual breeding but overlapping generations. *J. Genet.*, **58**, 122–4.

Haldane, J. B. S. & S. D. Jayakar. 1963. Polymorphism due to selection of varying direction. *J. Genet.*, **58**, 237–42.

Hamilton, W. D. 1966. The moulding of senescence by natural selection. *J. Theor. Biol.*, **12**, 12–45.

Harper, J. L. 1977. *Population Biology of Plants*. London, New York: Academic Press.

Harper, J. L. & J. White. 1974. The demography of plants. *Ann. Rev. Syst. Ecol.*, **5**, 419–63.

Hassell, M. P., J. H. Lawton & R. M. May. 1976. Patterns of dynamical behaviour in single species populations. *J. Anim. Ecol.*, **45**, 471–96.

Henry, L. 1953. Fondements théoriques des mesures de la fécondité naturelle. *Rev. Inst. Int. Stat.*, **21**, 135–51.

Henry, L. 1957. Fécondité et famille. Modéles mathématiques. I. *Population*, **12**, 413–44.

Henry, L. 1976. *Population. Analysis and Models*. London: Edward Arnold.

Hill, W. G. 1972. Effective size of populations with overlapping generations. *Theor. Popul. Biol.*, **3**, 278–89.

Hill, W. G. 1974. Prediction and evaluation of response to selection with overlapping generations. *Anim. Prod.*, **18**, 117–39.

Hill, W. G. 1977. Selection with overlapping generations. In *Proceedings of the International Conference on Quantitative Genetics*, ed. E. Pollak, O. Kempthorne and T. B. Bailey, pp. 367–78. Ames, Iowa: Iowa State University Press.

Hill, W. G. 1979. A note on effective population size with overlapping generations. *Genetics*, **92**, 317–22.

Hodder, V. M. 1963. Fecundity of Grand Bank haddock. *J. Fish. Res. Board Can.*, **20**, 1465–87.

Hoffman, G. W. 1974. On the origin of the genetic code and the stability of the translation apparatus. *J. Mol. Biol.*, **86**, 349–62.

Holgate, P. 1967. Population survival and life history phenomena. *J. Theor. Biol.*, **14**, 1–10.

Hoppensteadt, F. C. & J. B. Keller. 1976. Synchronization of periodical cicada emergences. *Science*, **194**, 335–7.

Jacquard, A. 1974. *The Genetic Structure of Populations*. Berlin: Springer–Verlag.

Jensen, L. 1973. Random selective advantage of genes and their probabilities of fixation. *Genet. Res.*, **21**, 215–19.

Johnson, D. L. 1977*a*. Variance-covariance structure of group means with overlapping generations. In *Proceedings of the International Conference on Quantitative Genetics*, ed. E. Pollak, O. Kempthorne and T. B. Bailey, pp. 851–8. Ames, Iowa: Iowa State University Press.

Johnson, D. L. 1977*b*. Inbreeding in populations with overlapping generations. *Genetics*, **87**, 581–91.

Karlin, S. & U. Lieberman. 1974. Random temporal variation in selection intensities: case of large population size. *Theor. Popul. Biol.*, **6**, 355–82.

Kempthorne, O. & E. Pollak. 1970. Concepts of fitness in Mendelian populations. *Genetics*, **64**, 125–45.

Keyfitz, N. 1968. *Introduction to the Mathematics of Population*. Reading, Mass.: Addison–Wesley.

Keyfitz, N. 1972. Population waves. In *Population Dynamics*, ed. T. N. E. Greville, pp. 1–38. New York, London: Academic Press.

Kim. Y. J. & Z. M. Sykes. 1976. An experimental study of weak ergodicity in human populations. *Theor. Popul. Biol.*, **10**, 150–72.

Kimura, M. 1955. Stochastic processes and distribution of gene frequencies under natural selection. *Cold Spring Harbor Symp. Quant. Biol.*, **20**, 33–53.

Kimura, M. 1968. Evolutionary rate at the molecular level. *Nature*, **217**, 624–6.

Kimura, M. 1978. Change of gene frequency by natural selection under population number regulation. *Proc. Natl. Acad. Sci. USA*, **75**, 1934–7.

Kimura, M. & J. F. Crow. 1963. The measurement of effective population number. *Evolution*, **17**, 279–88.

Kimura, M. & T. Ohta. 1971. *Theoretical Aspects of Population Genetics*. Princeton: Princeton University Press.

King, C. E. & W. W. Anderson. 1971. Age-specific selection. II. The interaction between r and K during population growth. *Am. Nat.*, **105**, 137–56.

Kirkwood, T. B. L. 1977. Evolution of ageing. *Nature*, **270**, 301–4.

Kirkwood, T. B. L. & R. Holliday. 1975. The stability of the translation apparatus. *J. Mol. Biol.*, **97**, 257–65.

Kirkwood, T. B. L. & R. Holliday. 1979. The evolution of ageing and longevity. *Proc. R. Soc. Lond.*, **B205**, 531–46.

Krebs, C. J. 1972. *Ecology*. New York: Harper & Row.

Lack, D. 1954. *The Natural Regulation of Animal Numbers*. Oxford: Clarendon Press.

Lack, D. 1966. *Population Studies of Birds*. Oxford: Clarendon Press.

Law, R. 1975. Colonisation and the evolution of life histories in *Poa annua*. Ph.D. thesis, University of Liverpool.

Law, R. 1979. The costs of reproduction in annual meadow grass. *Am. Nat.*, **113**, 3–16.

Law, R., A. D. Bradshaw & P. D. Putwain. 1977. Life-history variation in *Poa annua. Evolution*, **31**, 233–46.

Lawlor, L. R. 1976. Molting, growth and reproductive strategies in the terrestrial isopod *Armadillidium vulgare. Ecology*, **57**, 1179–94.

Leigh, E. G. 1970. Sex ratio and differential mortality between the sexes. *Am. Nat.*, **104**, 205–10.

Leigh, E. G., E. L. Charnov & R. R. Warner. 1976. Sex ratio, sex change, and natural selection. *Proc. Natl. Acad. Sci. USA*, **73**, 3656–60.

León, J. A. 1976. Life histories as adaptive strategies. *J. Theor. Biol.*, **60**, 301–36.

León, J. A. & B. Charlesworth. 1978. Ecological versions of Fisher's fundamental theorem of natural selection. *Ecology*, **59**, 457–64.

Leslie, P. H. 1945. On the use of matrices in certain population mathematics. *Biometrika*, **33**, 183–212.

Leslie, P. H. 1948. Some further remarks on the use of matrices in population mathematics. *Biometrika*, **35**, 213–45.

Levins, R. 1975. Evolution in communities near equilibrium. In *Ecology and Evolution of Communities*, ed. M. L. Cody and J. M. Diamond, pp. 16–50. Cambridge, Mass.: Belknap Press.

Lewis, E. G. 1942. On the generation and growth of a population. *Sankhya*, **6**, 93–96.

Lewontin, R. C. 1965. Selection for colonizing ability. In *The Genetics of Colonizing Species*, ed. H. G. Baker and G. L. Stebbins, pp. 77–94. New York, London: Academic Press.

Lloyd, M. & H. S. Dybas. 1966a. The periodical cicada problem. I. Population ecology. *Evolution*, **20**, 133–49.

Lloyd, M. & H. S. Dybas, 1966b. The periodical cicada problem. II. Evolution. *Evolution*, **20**, 466–505.

Lopez, A. 1961. *Problems in Stable Population Theory*. Princeton: Office of Population Research.

Lotka, A. J. 1925. *Elements of Physical Biology*. Baltimore: Williams & Watkins. (Reprinted and revised as *Elements of Mathematical Biology*, 1956. New York: Dover.)

MacArthur, R. H. 1962. Some generalized theorems of natural selection. *Proc. Natl. Acad. Sci. USA*, **38**, 1893–7.

MacArthur, R. H. & E. O. Wilson. 1967. *The Theory of Island Biogeography*. Princeton: Princeton University Press.

Malécot, G. 1948. *Les Mathématiques de l'Hérédité*. Paris: Masson.

May, R. M. 1973. *Stability and Complexity in Model Ecosystems*. Princeton: Princeton University Press.

May, R. M. & G. F. Oster. 1976. Bifurcations and dynamic complexity in simple ecological models. *Am. Nat.*, **110**, 573–99.

Maynard Smith, J. 1958. The effects of temperature and of egg-laying on the longevity of *Drosophila subobscura*. *J. Exp. Biol.*, **35**, 832–42.

Maynard Smith, J. 1964. Group selection and kin selection. *Nature*, **201**, 1145–7.

Maynard Smith, J. 1972. *On Evolution*. Edinburgh: Edinburgh University Press.

Maynard Smith, J. 1976. Group selection. *Q. Rev. Biol.*, **51**, 277–87.

Maynard Smith, J. & J. Haigh. 1974. The hitch-hiking effect of a favourable gene. *Genet. Res.*, **23**, 23–5.

Maynard Smith, J. & M. Slatkin. 1973. The stability of predator-prey systems. *Ecology*, **54**, 384–91.

Mayr, E. 1963. *Animal Species and Evolution*. Cambridge, Mass.: Belknap Press.

Medawar, P. B. 1946. Old age and natural death. *Modern Quarterly*, **1**, 30–56. (Reprinted in *The Uniqueness of the Individual* (1957), pp. 17–43. London: Methuen.)

Medawar, P. B. 1952. *An Unsolved Problem of Biology*. London: H. K. Lewis. (Reprinted in *The Uniqueness of the Individual* (1957), pp. 44–70. London: Methuen.)

Mertz, D. B. 1971*a*. The mathematical demography of the California condor population. *Am. Nat.*, **105**, 437–53.

Mertz, D. B. 1971*b*. Life history phenomena in increasing and decreasing populations. In *Statistical Ecology*, vol. 2. ed. G. P. Patil, E. C. Pielou and W. E. Waters, pp. 361–400. University Park, Pennsylvania: Pennsylvania State University Press.

Mertz, D. B. 1975. Senescent decline in flour beetles selected for early adult fitness. *Physiol. Zool.*, **48**, 1–23.

Michod, R. E. 1979. Evolution of life histories in response to age-specific mortality factors. *Am. Nat.*, **113**, 531–50.

Moran, P. A. P. 1950. Some remarks on animal population dynamics. *Biometrics*, **6**, 250–8.

Moran, P. A. P. 1962. *The Statistical Processes of Evolutionary Theory*. Oxford: Clarendon Press.

Mountford, M. D. 1971. Population survival in a variable environment. *J. Theor. Biol.*, **32**, 75–9.

Mountford, M. D. 1973. The significance of clutch-size. In *The Mathematical Theory of the Dynamics of Biological Populations*, ed. M. S. Bartlett and R. W. Hiorns, pp. 315–24. New York, London: Academic Press.

Murdoch, J. L., B. A. Walker, B. L. Halpern, J. W. Kuzma & V. A. McKusick. 1972. Life expectancy and causes of death in the Marfan syndrome. *New Engl. J. Med.*, **286**, 804–8.

Murphy, G. I. 1968. Pattern in life history phenomena and the environment. *Am. Nat.*, **102**, 52–64.

Nagylaki, T. 1976. The evolution of one- and two-locus systems. *Genetics*, **83**, 583–600.

Nagylaki, T. 1977. *Selection in One- and Two-Locus Systems*. Berlin: Springer–Verlag.

Nagylaki, T. 1979*a*. The dynamics of density- and frequency-dependent selection. *Proc. Natl. Acad. Sci. USA*, **76**, 483–41.

Nagylaki, T. 1979*b*. Selection in dioecious populations. *Ann. Hum. Genet.*, **43**, 141–8.

Nagylaki, T. & J. F. Crow. 1974. Continuous selective models. *Theor. Popul. Biol.*, **5**, 257–83.

Nei, M. 1970. Effective size of human populations. *Am. J. Hum. Genet.*, **22**, 694–6.

Nei, M. 1975. *Molecular Population Genetics and Evolution*. Amsterdam: North-Holland.

Nei, M. & Y. Imaizumi. 1966. Genetic structure of human populations.

II. Differentiation of blood group frequencies among isolated populations. *Heredity*, **21**, 183–90.

Nelson, J. P. 1969. The breeding ecology of the red-footed booby in the Galapagos. *J. Anim. Ecol.*, **38**, 181–98.

Nicholson, A. J. 1957. The self-adjustment of populations to change. *Cold Spring Harbor Symp. Quant. Biol.*, **22**, 153–73.

Nicholson, A. J. 1960. The role of population dynamics in natural selection. In *Evolution after Darwin*, vol. 1, *The Evolution of Life*, ed. S. Tax, pp. 477–520. Chicago: University of Chicago Press.

Norton, H. T. J. 1928. Natural selection and Mendelian variation. *Proc. Lond. Math. Soc.*, **28**, 1–45.

Novitski, E. & A. W. Kimball. 1958. Birth order, parental ages and sex of offspring. *Am. J. Hum. Genet.*, **21**, 123–31.

Orgel, L. E. 1963. The maintenance of the accuracy of protein synthesis and its relevance to ageing. *Proc. Natl. Acad. Sci. USA*, **49**, 517–21.

Orgel, L. E. 1973. Ageing of clones of mammalian cells. *Nature*, **243**, 441–5.

Orians, G. H. 1969. Age and hunting success in the brown pelican (*Pelecanus occidentalis*). *Anim. Behav.*, **17**, 316–19.

Oster, G. F. 1976. Internal variables in population dynamics. In *Lectures on Mathematics in the Life Sciences*, vol. 8, ed. S. Levin, pp. 37–68. Providence, R.I.: American Mathematical Society.

Oster, G. F. & Y. Takahashi. 1974. Models for age-specific interactions in a periodic environment. *Ecol. Monogr.*, **44**, 483–501.

Parlett, B. 1972. Can there be a marriage function? In *Population Dynamics*, ed. T. N. E. Greville, pp. 107–135. New York, London: Academic Press.

Pennycuick, C. J., R. M. Compton & L. Beckingham 1968. A computer model for simulating the growth of a population or two populations. *J. Theor. Biol.*, **18**, 316–29.

Perrins, C. M. & D. Moss. 1974. Survival of young great tits in relation to age of female parent. *Ibis*, **116**, 220–4.

Pianka, E. R. 1970. On *r* and *K* selection. *Am. Nat.*, **104**, 592–6.

Pianka, E. R. 1974. *Evolutionary Ecology*. New York: Harper & Row.

Pianka, E. R. & W. S. Parker. 1975. Age-specific reproductive tactics. *Am. Nat.*, **109**, 453–63.

Pollak, E. 1976. A stochastic treatment for rare genes in large populations with overlapping generations. *Theor. Popul. Biol.*, **10**, 109–26.

Pollak, E. 1977. Selective advance in populations with overlapping generations. In *Proceedings of the International Conference on Quantitative Genetics*, ed. E. Pollak, O. Kempthorne and T. B. Bailey, pp. 379–97. Ames, Iowa: Iowa State University Press.

Pollak, E. & O. Kempthorne. 1970. Malthusian parameters in genetic populations. Part I. Haploid and selfing models. *Theor. Popul. Biol.*, **1**, 315–45.

Pollak, E. & O. Kempthorne. 1971. Malthusian parameters in genetic populations. Part II. Random mating populations in infinite habitats. *Theor. Popul. Biol.*, **2**, 351–90.

Pollard, J. H. 1973. *Mathematical Models for the Growth of Human Populations*. Cambridge University Press.

Price, G. R. 1972. Fisher's Fundamental Theorem made clear. *Ann. Hum. Genet.*, **36**, 129–40.

Price, G. R. & C. A. B. Smith. 1972. Fisher's Malthusian parameter and reproductive value. *Ann. Hum. Genet.*, **36**, 1-7.

Prout, T. 1968. Sufficient conditions for multiple niche polymorphisms. *Am. Nat.*, **102**, 493-6.

Prout, T. 1971*a*. The relation between fitness components and population prediction in *Drosophila*. I. The estimation of fitness components. *Genetics*, **68**, 127-49.

Prout, T. 1971*b*. The relation between fitness components and population prediction in *Drosophila*. II. Population prediction. *Genetics*, **68**, 151-67.

Reed, T. E. & J. H. Chandler. 1958. Huntington's chorea in Michigan. I. Demography and genetics. *Am. J. Hum. Genet.*, **10**, 201-21.

Reed, T. E. & J. V. Neel. 1955. A genetic study of multiple polyposis of the colon. *Am. J. Hum. Genet.*, **7**, 236-63.

Reed, T. E. & J. V. Neel. 1959. Huntington's chorea in Michigan. 2. Selection and mutation. *Am. J. Hum. Genet.*, **11**, 107-36.

Rendel, J. M. & A. Robertson. 1950. Estimation of genetic gain in milk yield by selection in a closed herd of dairy cattle. *J. Genet.*, **50**, 1-8.

Rice, D. W. & K. W. Kenyon. 1962. Breeding cycles and the behaviour of Laysan and Black-footed Albatrosses. *Auk*, **79**, 517-67.

Ricker, W. E. 1954. Stock and recruitment. *J. Fish. Res. Board Can.*, **1**, 559-623.

Ricklefs, R. E. 1973. *Ecology*. Newton, Mass: Chiron Press.

Ricklefs, R. E. 1977. On the evolution of reproductive strategies in birds: reproductive effort. *Am. Nat.*, **111**, 453-78.

Robbins, R. B. 1918. Some applications of mathematics to breeding problems. II. *Genetics*, **3**, 73-92.

Rorres, C. 1976. Stability of an age-specific population with density-dependent fertility. *Theor. Popul. Biol.*, **10**, 26-46.

Rose, M. R. 1979. Quantitative genetics of adult female life-history in *Drosophila melanogaster*. D.Phil thesis, University of Sussex.

Roughgarden, J. 1971. Density-dependent natural selection. *Ecology*, **52**, 453-68.

Roughgarden, J. 1976. Resource partitioning among competing species – a coevolutionary approach. *Theor. Popul. Biol.*, **9**, 388-424.

Sacher, G. A. 1959. Relationship of lifespan to brain weight and weight in animals. In *The Life-span of Animals. CIBA Foundation Colloquia on Ageing*, vol. 5, ed. G. E. Wolstenholme and M. O'Connor, pp. 115-33. London: Churchill.

Sacher, G. A. 1978. Evolution of longevity and survival characteristics. In *The Genetics of Ageing*, ed. E. L. Schneider, pp. 151-68. New York: Plenum Press.

Sarukhan, J. & M. Gadgil. 1974. Studies on plant demography: *Ranunculus repens* L., *R. bulbosus* L. and *R. acris*. L. III. A mathematical model incorporating multiple modes of reproduction. *J. Ecol.*, **62**, 921-36.

Schaal, B. A. & D. A. Levin. 1976. Demographic genetics of *Liatris cylindracea* Michx. (Compositae). *Am. Nat.*, **110**, 191-206.

Schaffer, W. M. 1974*a*. Selection for optimal life histories: the effects of age structure. *Ecology*, **55**, 291-303.

Schaffer, W. M. 1974*b*. Optimal reproductive effort in fluctuating environments. *Am. Nat.*, **108**, 783-90.

Schaffer, W. M. & M. Gadgil. 1975. Selection for optimal life histories in plants. In *Ecology and Evolution of Communities*, ed. M. Cody and J. E. Diamond, pp. 142–57. Cambridge, Mass.: Belknap Press.

Schaffer, W. M. & M. L. Rosenzweig. 1977. Selection for optimal life histories. II. Multiple equilibria and the evolution of alternative reproductive strategies. *Ecology*, **58**, 60–72.

Schaffer, W. M. & M. V. Schaffer. 1977. The adaptive significance of variations in reproductive habit in the Agavaceae. In *Evolutionary Ecology*, ed. B. Stonehouse and C. M. Perrins, pp. 261–76. London: Macmillan.

Schaffer, W. M. & M. V. Schaffer. 1979. The adaptive significance of variations in reproductive habit in the Agavaceae. II. Pollinator foraging behaviour and selection for increased reproductive expenditure. *Ecology*, in press.

Schneider, E. L. 1978. *The Genetics of Ageing*. New York: Plenum Press.

Selander, R. K. 1965. On mating systems and sexual selection. *Am. Nat.*, **99**, 129–41.

Seneta, E. 1973. *Non-negative Matrices*. London: Allen & Unwin.

Sharpe, F. R. & A. J. Lotka. 1911. A problem in age distribution. *Philos. Mag.*, **21**, 435–8.

Simmons, M. J. & J. F. Crow. 1977. Mutations affecting fitness in *Drosophila* populations. *Ann. Rev. Genet.*, **11**, 49–78.

Slobodkin, L. 1974. Prudent predation does not require group selection. *Am. Nat.*, **108**, 665–78.

Smouse, P. E. & K. M. Weiss. 1975. Discrete demographic models with density-dependent vital rates. *Oecologia*, **21**, 205–18.

Snell, T. W. & C. E. King. 1977. Lifespan and fecundity patterns in rotifers: the cost of reproduction. *Evolution*, **31**, 882–90.

Sokal, R. R. 1970. Senescence and genetic load: evidence from *Tribolium*. *Science*, **167**, 1733–4.

Solbrig, O. T. & B. B. Simpson. 1974. Components of regulation of a population of dandelions in Michigan. *J. Ecol.*, **62**, 473–86.

Southwood, T. R. E. 1966. *Ecological Methods*. London: Methuen.

Southwood, T. R. E., R. M. May, M. P. Hassell & G. R. Conway. 1974. Ecological strategies and population parameters. *Am. Nat.*, **108**, 791–804.

Spieth, P. T. 1974. Theoretical considerations of unequal sex ratios. *Am. Nat.*, **108**, 837–49.

Stearns, S. C. 1976. Life history tactics: a review of the ideas. *Q. Rev. Biol.*, **51**, 3–47.

Stearns, S. C. 1977. The evolution of life history traits: a critique of the theory and a review of the data. *Ann. Rev. Ecol. Syst.*, **8**, 145–71.

Stebbins, G. L. 1950. *Variation and Evolution in Plants*. New York: Columbia University Press.

Storer, J. B. 1966. Longevity and gross pathology at death in 22 inbred mouse strains. *J. Gerontol.*, **21**, 404–9.

Tamarin, R. H. & C. J. Krebs. 1969. *Microtus* population biology. II. Genetic changes at the transferrin locus in fluctuating populations of two vole species. *Evolution*, **23**, 183–211.

Taylor, H. M., R. S. Gourley & C. E. Lawrence. 1974. Natural selection of life history attributes: an analytical approach. *Theor. Popul. Biol.*, **5**, 104–22.

Teitelbaum, M. S. 1972. Factors associated with the sex ratio in human populations. In *The Structure of Human Populations*, ed. G. A. Harrison and A. J. Boyce, pp. 90–109. Oxford: Clarendon Press.

Templeton, A. R. & D. A. Levin. 1979. Evolutionary consequences of seed pools. *Am. Nat.*, **114**, 232–49.

Tinkle, D. W. & N. F. Hadley. 1973. Reproductive effort and winter activity in the viviparous montane lizard *Sceloporous jarrovi*. *Copeia*, **1973**, 272–7.

Tinkle, D. W. & N. F. Hadley. 1975. Lizard reproductive effort: caloric estimates and comments on its evolution. *Ecology*, **56**, 427–34.

Trivers, R. L. & H. Hare. 1976. Haplodiploidy and the evolution of the social insects. *Science*, **191**, 249–63.

Turner, J. R. G. 1970. Changes in mean fitness under natural selection. In *Mathematical Topics in Population Genetics*, ed. K. Kojima, pp. 32–78. Berlin: Springer-Verlag.

Usher, M. B. 1972. Developments in the Leslie matrix model. In *Mathematical Models in Ecology*, ed. J. N. R. Jeffers, pp. 29–60. Oxford: Blackwell.

Warner, R. R. 1975. The adaptive significance of sequential hermaphroditism in animals. *Am. Nat.*, **109**, 61–82.

Warner, R. R., D. R. Robertson & E. G. Leigh. 1975. Sex change and sexual selection. *Science*, **190**, 633–8.

Watanabe, T., W. W. Anderson, T. Dobzhansky & O. Pavlovsky. 1970. Selection in experimental populations of *Drosophila pseudoobscura* with different initial frequencies of chromosomal variants. *Genet. Res.*, **15**, 123–9.

Wilbur, H. M., D. W. Tinkle & J. P. Collins. 1974. Environmental certainty, trophic level and resource availability in life history evolution. *Amer. Nat.*, **108**, 805–17.

Wiley, R. H. 1974*a*. Evolution of social organization and life-history patterns among grouse. *Q. Rev. Biol.*, **49**, 201–27.

Wiley, R. H. 1974*b*. Effects of delayed reproduction on survival, fecundity, and the rate of population increase. *Am. Nat.*, **108**, 705–9.

Williams, G. C. 1957. Pleiotropy, natural selection and the evolution of senescence. *Evolution*, **11**, 398–411.

Williams, G. C. 1966*a*. *Adaptation and Natural Selection*. Princeton: Princeton University Press.

Williams, G. C. 1966*b*. Natural selection, the costs of reproduction and a refinement of Lack's principle. *Am. Nat.*, **100**, 687–90.

Williamson, J. A. & B. Charlesworth. 1976. The effect of age of founder on the probability of survival of a colony. *J. Theor. Biol.*, **56**, 175–90.

Wilson, E. O. & W. H. Bossert. 1971. *A Primer of Population Biology*. Stamford, Connecticut: Sinauer.

Wright, S. 1929*a*. Fisher's theory of dominance. *Am. Nat.*, **63**, 274–9.

Wright, S. 1929*b*. The evolution of dominance. Comment on Dr. Fisher's reply. *Am. Nat.*, **63**, 556–61.

Wright, S. 1931. Evolution in Mendelian populations. *Genetics*, **16**, 97–159.

Wright, S. 1938. Size of population and breeding structure in relation to evolution. *Science*, **87**, 430–1.

Wright, S. 1942. Statistical genetics and evolution. *Bull. Am. Math. Soc.*, **48**, 223–46.

Wynne-Edwards, V. C. 1955. Low reproductive rates in birds, especially sea-birds. *Proc. Int. Ornithol. Congr.*, **11**, 540–7.

Wynne-Edwards, V. C. 1962. *Animal Dispersal in Relation to Social Behaviour*. Edinburgh: Oliver & Boyd.

AUTHOR INDEX

SUBJECT INDEX